More information about this series at http://www.springer.com/series/8059

Kyle Konis · Stephen Selkowitz

Effective Daylighting with High-Performance Facades

Emerging Design Practices

 Springer

Kyle Konis
School of Architecture
University of Southern California
Los Angeles, CA
USA

Stephen Selkowitz
Lawrence Berkeley National Laboratory
Berkeley, CA
USA

ISSN 1865-3529   ISSN 1865-3537 (electronic)
Green Energy and Technology
ISBN 978-3-319-39461-9   ISBN 978-3-319-39463-3 (eBook)
DOI 10.1007/978-3-319-39463-3

Library of Congress Control Number: 2017932534

Printed on acid-free paper

This Springer imprint is published by Springer Nature
The registered company is Springer International Publishing AG
The registered company address is: Gewerbestrasse 11, 6330 Cham, Switzerland

Preface

Our planet has been bathed in daylight from the sun for over 4 billion years. The role of daylight in buildings has profoundly changed over the time frame of the last century. In the early twentieth century, electric light began to displace what had been the dominant source of lumens in a building interior for all of human existence —the sun and sky. Due to rapid improvements in electric lighting technology, e.g., fluorescent lamps, and the ubiquity of the electric grid with low cost power, electric lighting became the dominant luminous source in buildings by mid-century. The oil crisis of the 1970s followed by a new focus on the environmental impacts of carbon emissions and sustainability challenges at the end of the twentieth century refocused new attention on effective daylighted building designs. But by the first two decades of the twenty-first century, new advances in electric lighting and smart controls have once again reopened the discussion and debate on the appropriate role of daylight in buildings.

This book captures the result of multiyear efforts by the authors to address those issues by postulating and exploring a dual pathway whereby effective daylight design continues to serve as a powerful energy management strategy while enhancing its value to occupants for visual performance, view, comfort, and health. This leads to exploring the technical potential of the building facade as a mechanism for utilizing environmental services provided by natural systems to address building energy and carbon reduction goals, while enhancing occupant experience and well-being. Daylight is a renewable source of high efficacy light, which makes the daylighting of buildings an attractive energy design strategy compared to standard practices of electrical lighting design. And there is a growing body of scientific knowledge linking access to sufficient daylight and window views with improved health and well-being. While the daylighting of buildings is not a new topic, one of the central barriers to effective daylighting is that daylighting *performance* is often narrowly defined by different stakeholders, leading to a fragmented approach to performance assessment in the design and operational life-cycle of buildings. Despite decades of discussion and design, there remain relatively few successful examples of projects where performance outcomes consistently meet design intent, particularly from the perspective of building occupants and measured reductions in energy use.

Daylight cannot be easily separated from electric lighting in its impact on people and buildings. It is unique as a light source in terms of its intrinsic variability (intensity, directionality and color) over time and weather, and the differences one can experience spatially within a building. Furthermore, at a time when the efficacy and control of electric lighting is improving, after decades without major change, the continued importance of daylight design has been questioned. This book explores the case for advanced building-facade daylighting design practices informed both by important energy, power, and carbon emissions constraints as well as by human-centric factors such as health, comfort, and performance. The state-of-the-art approaches are discussed in the context of simulation-based design workflows, innovative technologies, and real project case studies, all targeting low and Zero Net Energy (ZNE) solutions. The book seeks to redefine effective daylighting by challenging the contemporary approach to glazing and facade system design. Contemporary design is often driven by the goal of architectural "transparency," pursued through the near-universal application of a sealed and static, highly glazed building skin to projects across the globe. While "transparent" facades have become one of the most iconic symbols for buildings promoted as "sustainable," "green," or "high-performance," these designs often fail to achieve claimed energy savings and can be visually and thermally uncomfortable. The book argues that we must replace this simplified approach to design and engineering with alternate approaches that more effectively incorporate local site and climate, carbon reduction goals, and the needs of building occupants as critical drivers of building performance, design solutions, and technological innovation.

While the book is informed by a broad spectrum of work by researchers and designers over the last 50 years, it focuses on the recent evolution of technology, systems and software solutions that are changing how buildings are designed and operated today, and explores how that might evolve in the future. In comparing "what is" to "what is needed" the book suggests the need to shift design practices from:

- The application of universal design guidance to climate, people and program specific design goals.
- Static, unresponsive systems to dynamic, adaptive systems.
- Homogeneous generic indoor work environments to granular, personalized environments.
- Fragmented collections of building components towards integrated (and interconnected) daylighting/perimeter-zone systems.
- Rule-of-thumb design guidance to evidence-based design solutions.
- Compliance-based prescriptive workflows to performance-based design workflows.

Broader principles and trends are illustrated with examples from the authors' studies and from the design community at large. Readers benefit from a comprehensive approach that addresses the world of design and engineering through a focus on building occupants. The book is intended for architects, lighting designers, facade engineers, manufacturers, building owners/operators, and advanced students, all of whom are essential partners in the drive to capture the full benefits that effective daylight design offers for people, for buildings and for the environment.

Los Angeles, USA   Kyle Konis
Berkeley, USA   Stephen Selkowitz

Acknowledgements

It is with great pleasure that we acknowledge the contribution of a broad array of groups and individuals in the development of the ideas and materials within this book.

The writing of this book was supported in part by the University of Southern California School of Architecture, which provided infrastructure, graduate research assistant support and financial support to Prof. Kyle Konis. As a faculty member teaching in the Chase L. Leavitt Graduate Building Science Program, Prof. Konis has benefitted from a program with a strong and sustained focus on environmentally responsive design practices, particularly the work of Profs. Ralph Knowles and Marc Schiler in passive solar architecture. Professor Konis has also benefitted from a supportive group of faculty colleagues and a capable body of Master of Building Science (MBS) students. Notably, the form-finding workflow example presented in Chap. 4 (section 4.2.2) began as the thesis work of Alejandro Gamas (under the supervision of Profs. Kyle Konis, Karen Kensek, and Douglas Noble). In addition, the thesis work of MBS student Kelly Burkhart, which utilized the cart-mounted spectrometer and High Dynamic Range enabled camera discussed in Chap. 2 (section 2.3.2) represents a pioneering effort to spatially map the circadian effectiveness of real daylighted buildings in use. USC graduate research assistants Sue Long Lee and Yang Li both contributed to architectural graphics used in the book. Prof. Konis would also like to acknowledge the training and mentorship of staff scientist Eleanor S. Lee, who served as a supervisor to Kyle when he worked as a graduate research assistant at the Lawrence Berkeley National Laboratory (LBNL). Finally, Prof. Konis would like to acknowledge the influential work of U.C. Berkeley Profs. Edward Arens, Gail Brager, Charles Benton, and Susan Ubbelohde, who have all demonstrated the importance of design from the perspective of end-users through their research activities, tools development, and design practices.

The book draws heavily on tools, data, and reports authored by Stephen Selkowitz and a team of colleagues in the Windows and Envelope Materials Group in the Building Technology and Urban Systems Division at LBNL. Initiated by

Selkowitz in 1976 as the "Windows and Daylighting Group," that team of about 20 researchers has collaborated over a 40 year period with numerous public and private partners (fellow researchers, architects, engineers, manufacturers), in the U.S. and globally, in the development of new technologies, systems, and tools to enhance the performance and effectiveness not only of daylighted buildings, but also for a wide range of high-performance coatings, glazings, and facade systems as well. That body of work was facilitated by the creation of an extensive and unique physical testing infrastructure at LBNL and a suite of daylight simulation tools (Superlite, Radiance, DOE-2, EnergyPlus, WINDOW, COMFEN). The long-term RD&D effort was made possible by continuous financial support from the U.S. Department of Energy, Building Technologies Office, and with significant additional project support over many years from the California Energy Commission, U.S. General Services Administration, New York State ERDA, and numerous other public and private partners and collaborators. Over 200 journal articles, research reports and conference proceedings were published to share results with the broader daylighting community. The names of all the contributing LBNL windows and daylighting research staff and visiting researchers from other global centers of daylighting research is too large to list (see authorship of papers on the LBNL website https:// eta.lbl.gov/publications) but the daylighting research thrust of the team has been ably led by Eleanor Lee over a 25-year period.

The numerous examples of effective daylighting applied in practice would not have been possible without the support of researchers, practicing designers and the willingness of principles at various firms to share project information, particularly performance data. We thank staff at Behnisch Architekten (Michael Kocher and Robert Matthew Noblett), Transsolar KlimaEngineering (Thomas Auer), the Lawrence Berkeley National Laboratory (Eleanor Lee, Andy McNeil), the National Renewable Energy Laboratory (Rob Guglielmetti and Jennifer Scheib), RNL (Mundi Wahlberg) ARUP (Galen Burrell), The Miller Hull Partnership (Wendy Abeel), Perkins+Will (Andrew TsayJacobs), the University of Washington (Chris Meek and Robert Peña), Henning Larsen Architects (Anne Iversen, Micki Aaen Petersen and Jakob Strømann-Andersen), and U.C. Berkeley (Prof. Charles Benton) for providing information on the projects presented in the book.

Additionally, the authors would like to thank a variety of companies for providing information on their facade and daylighting technologies: Okalux, Panelite, Microshade, SunCentral, View, and Sage Glass. As products evolve continuously, the reader is encouraged to visit the websites of the manufacturers for current product data and availability. Given the limited scope of any book, it is never possible to reference all the related technologies or manufacturers with relevant products nor to comment in detail on the generalized performance of any design solution for applications beyond those discussed in these chapters.

Finally, we are grateful to our families, who support our efforts with encouragement, thoughtful criticism, and tolerance of our work schedules.

Contents

Abbreviations

AEC	Architecture, Engineering and Construction
AIA	American Institute of Architects
ASGS	Angular Selective Glazing Systems
ASHRAE	American Society of Heating Refrigeration and Air-conditioning Engineers
BAS	Building Automation Systems
BAU	Business As Usual
BIM	Building Information Modeling
BIPV	Building Integrated Photovoltaics
BSDF	Bidirectional Scattering Distribution Function
Btu	British Thermal Unit
CAD	Computer-Aided Design
CBD	Commercial Building Disclosure
CBDM	Climate Based Daylight Modeling
CDD	Cooling Degree Days
CEC	California Energy Commission
CEUS	California Energy End Use Survey
CGDB	Complex Glazing and Shading Database
CIE	Commission Internationale de l'Eclairage
COMFEN	Commercial Fenestration
COP	Coefficient of Performance
DDGE	Dynamic Daylight Glare Evaluation
DGI	Daylight Glare Index
DGP	Daylight Glare Probability
DOE	Department of Energy (U.S.)
EC	Electrochromic
EIA	Energy Information Administration (U.S.)
EUI	Energy Use Intensity
GHG	Greenhouse Gas
GIS	Geographic Information System

HDR	High Dynamic Range
HVAC	Heating Ventilation and Air Conditioning
IBL	Image-Based Lighting
IEA	International Energy Agency
IEQ	Indoor Environmental Quality
IGDB	International Glazing Database
IGU	Insulating Glazing Unit
IoT	Internet of Things
IoTePS	Internet-of-Things-enabled Perimeter Systems
LBNL	Lawrence Berkeley National Laboratory
LED	Light Emitting Diode
LEED	Leadership in Energy and Environmental Design
LSG	Light-to-Solar-Gain ratio
Lux	Photometric unit of illuminance
MRT	Mean Radiant Temperature
NYT	New York Times
OEM	Original Equipment Manufacturer
OLS	Optical Light Redirecting System
PMP	Performance Measurement Protocols
POE	Post Occupancy Evaluation
PV	Photovoltaics
RD&D	Research Development and Deployment
RGB	Red, Green, Blue
RSF	Research Support Facility
SC	Shading Coefficient
SHGC	Solar Heat Gain Coefficient
SSL	Solid-State Lighting
UFAD	Underfloor Air Distribution System
VLT	Visible Light Transmittance
WWR	Window-to-Wall Ratio
ZNE	Zero Net Energy

Chapter 1
The Challenge of Effective Daylighting

1.1 Introduction

Effective use of daylight in buildings is a fundamental consideration for minimizing the carbon impacts of the built environment and for creating indoor environments that support the comfort, performance and well-being of building occupants. Highly glazed, "transparent" facades have become iconic images for buildings promoted as "sustainable," "green," or "high-performance," but these designs often fail to capture the claimed energy savings and may be thermally and visually uncomfortable. Little guidance exists for designers to examine how human-factors objectives such as daylight sufficiency, visual comfort and view should be defined, measured, and evaluated in context with whole-building energy objectives to establish confidence that goals for project performance can be realized after value engineering, construction, commissioning and occupancy. The integration of facade technologies, controls, and other building systems with occupant needs and the reality of building operations is a complex task, which requires a comprehensive and continuous approach. This book argues that effective daylighting requires the development of strategies and methods that acknowledge the needs and behaviors of building occupants as a critical determinant of long-term energy performance. The book defines effective daylighting with specific energy and human-factors performance objectives. It presents a range of promising daylighting design strategies and discusses them in context with simulation-based workflows and project case studies. Finally, the book presents and discusses the ongoing evolution of the glazing, shading and light control technologies and systems that underlie daylight solutions, and the applicability of emerging methodologies for optimizing and validating daylighting performance.

The following sections outline the key challenges to effective use of daylight in the design and operation of high-performance buildings to reduce carbon impacts and enhance the quality of the indoor environment for building occupants. The chapter concludes by introducing an agenda to address these issues at scale, consisting of three central transformations to contemporary design practices:

© Springer International Publishing Switzerland 2017
K. Konis and S. Selkowitz, *Effective Daylighting with High-Performance Facades*, Green Energy and Technology, DOI 10.1007/978-3-319-39463-3_1

1. From compliance-based to performance-based design.
2. From static and unresponsive to context-aware and adaptive systems.
3. From theory to validation, feedback and learning.

1.2 Effective Daylighting as a Central Driver for Low-Energy, Low-Carbon Buildings

The design of new high-performance buildings and the application of deep-energy retrofits to existing buildings will play a key role in the development of a low-carbon future. There is broad agreement that aggressive greenhouse gas (GHG) mitigation strategies are needed in order to maintain atmospheric CO_2 emissions below 450 ppm, and limit global equilibrium temperature rise to 2 °C above preindustrial levels, the threshold considered critical for avoiding irreversible effects of climate. Buildings account for more than 32% of total global energy consumption and one third of global black carbon emissions, primarily through the use of fossil fuels[1] during their operational life-cycle (Lucon et al. 2014). Looking ahead, global building energy consumption is predicted to double or even triple by 2050 as the global population increases and more consumers in the developing world gain access to energy-intensive modern buildings and operational practices. Of all sectors, (energy supply, transport, buildings, industry, agriculture, forestry, and waste), buildings have the greatest economic potential[2] for mitigation through the whole-building integration of environmentally responsive design strategies, low-energy building systems, and greater levels of energy-awareness and engagement from building occupants. A global vision that drives the existing and new building stock toward Zero Net Energy (ZNE), or even net positive performance levels would profoundly change the environmental impact of the building sector on our planet.

In commercial buildings, which account for roughly half of the energy used by all U.S. buildings (U.S. DOE 2011), decisions related to fenestration in the building envelope directly affect the largest energy end uses (HVAC and lighting) and are thus a central area of focus for performance improvements aimed at enabling low energy buildings. Replacing one square meter of opaque building envelope with a transparent element causes three fundamental changes to the energy balance of a building: (1) it admits daylight which can be used to offset electric lighting use, (2) it increases direct conductive/convective thermal losses/gains that can increase heating and cooling loads, and (3) it increases solar gain which might offset heating in winter but increase cooling in summer. Given the range of building types, sizes, and climates there is wide variability case to case. But in most instances these design decisions have significant impact on overall building loads and resultant energy use, as well as occupant comfort.

---

[1]Most of building GHG emissions (6.02 Gt of 9.18 GtCO$_2$eq) are indirect CO_2 emissions from the consumption of electricity.

[2]https://www.ipcc.ch/publications_and_data/ar4/syr/en/mains4-3.html (Fig. 4.2, WGIII Fig. SPM.6).

Daylight is a renewable source of high efficacy light, which makes the daylighting of buildings an attractive energy strategy compared to the standard practice of continuous electrical lighting. In the United States, lighting represents the single largest commercial building electricity end use (0.78 exajoules (EJ)) (724 Trillion Btu) (EIA 2012), and is consumed primarily during daylight hours. Of the total averages, it is estimated that 60% is consumed in perimeter zones[3] located 0–12.2 m (0–40 ft) from the building facade during typical daytime work hours (8:00–18:00) (Shehabi et al. 2013). One square meter of sunlight contains enough lumens to illuminate 200 m^2 of floor space, so the challenge is control and distribution. The luminous efficacy of daylight, as filtered through spectrally selective glazing is also good, in the same range as the best available LED lamp efficacy or $\sim$ 120–250 lumens/watt. The key challenge of daylight is to distribute it effectively across the occupied floor area and to control glare from both sun and sky. Diffuse daylight (directly from the sky, reflected from exterior surfaces or diffused from sun control systems) can provide adequate flux to reduce electric lighting to a five-meter depth in an office. Redirecting sunlight via active and passive daylighting optics can extend that range to over ten meters.

Cooling loads represent another significant energy end use (14%), and one-third is due to electrical lighting and another one-third to solar heat gains through windows (Huang and Franconi 1999). Because low-energy projects often implement passive or low-energy cooling alternatives such as radiant systems or exposed thermal mass with night-flush ventilation, effective solar control is a requirement to avoid exceeding the cooling capacities of these systems, which are typically lower than conventional mechanical HVAC, and consequently more sensitive to peak solar heat gains. Therefore, fenestration strategies that control solar loads while transmitting sufficient daylight to minimize the need for electrical lighting in perimeter zones have the potential to significantly improve overall energy performance.

The goal in achieving dramatic reductions in building energy use is to convert building facades from their current role as a net energy cost to a net benefit. This requires converting the facade from a net energy loser to energy neutral, or even a supplier of energy on an annual basis by reducing thermal losses, actively managing thermal gains, integrating operable windows to reduce cooling and ventilation loads, utilizing daylight to offset electric lighting and integrating solar collection (e.g. solar photovoltaic or transpired solar collector systems). For example, the Lawrence Berkeley National Laboratory (LBNL) has calculated that total window area in the U.S. commercial building stock currently consumes 1.56 (EJ) but could be converted to a 1.16 EJ net energy gain if all windows were converted to high performance systems (Apte et al. 2006) (Fig. 1.1). Simulation-based studies using a standard office building located in Chicago, IL have shown that even the application of available, "off-the-shelf" fenestration technology packages can outperform opaque insulated walls (Lee et al. 2009) if intelligently designed and managed.

However, recasting the building envelope as a supportive element of the building energy concept represents a more complex challenge than a simple technology switch. Effectively utilizing the building envelope as a mechanism to

---

[3]Excluding non-applicable floor space such as religious worship or vacant space.

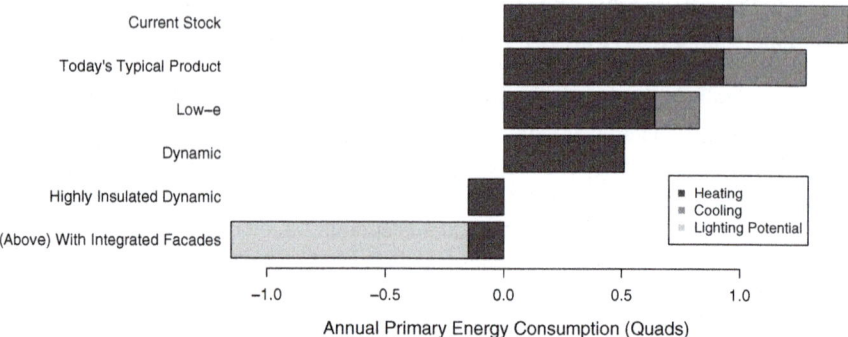

Fig. 1.1 Facade energy impacts in U.S. Commercial Building Sector (data from Apte et al. 2006). Current facade/window stock is estimated to consume ~ 1.56 EJ (~ 1.48 Quadrillion Btu (Quads)) (~ $20B USD); replacement by improved technologies reduces energy as indicated; "Integrated Facades" with full daylight potential offsets lighting loads of 1.05 EJ (1.0 Quads) for a total net reduction of 1.20 EJ (1.14 Quads)

leverage environmental services available from natural systems requires fundamental changes to contemporary (i.e. Business As Usual (BAU)) design practices, particularly in regard to building form, massing, and interior organization. As one notable example, the John & Frances Angelos Law Center demonstrates the integration of building form, facade elements and building systems to minimize demand for mechanical space conditioning and electrical lighting energy in a large 17,837 m^2 (192,000 ft^2) academic building. Located in a cooling-dominated climate (Baltimore, MD), where sealed facades and air-conditioning are standard practice, the project illustrates one case study of an environmentally-responsive alternative model, which yields additional co-benefits for building occupants through the provision of greater access to daylight, visual connection to the exterior, and greater control over indoor environmental conditions.

The Law Center program is subdivided into individual volumes (Fig. 1.2), which interlock with a multi-story daylit atrium (Fig. 1.3). The void space created

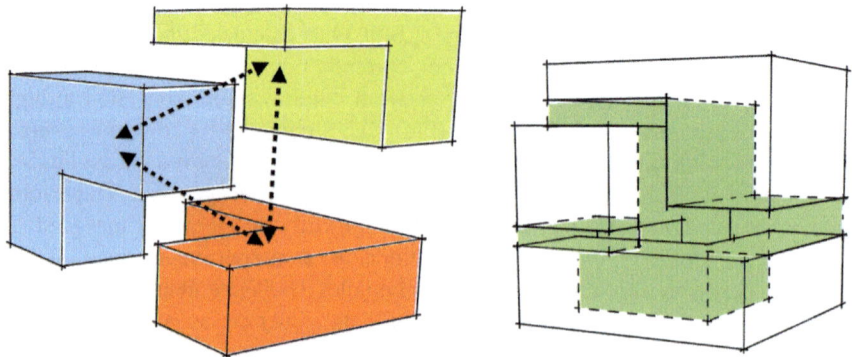

Fig. 1.2 Subdivision of the John & Frances Angelos Law Center program into individual volumes, which interlock with a multi-story daylit atrium. *Image credit* Behnisch Architekten

Fig. 1.3 Building section of the John & Frances Angelos Law Center showing the side and top-lit daylit atrium space that serves as the primary means of circulation. *Image credit* Behnisch Architekten

by the separation of program volumes increases the available surface area for fenestration, both on the exterior building envelope and facing the interior court-yard, enabling all regularly occupied areas of the building to have access to daylight and views, while simultaneously seeking to achieve a low whole-building energy target through the utilization of the building envelope (Fig. 1.4) for daylighting, management of solar gains, natural ventilation and space cooling. Massing and envelope strategies are supplemented with dynamic (climate responsive) facade solar shading, automated windows, thermally active interior surfaces, and

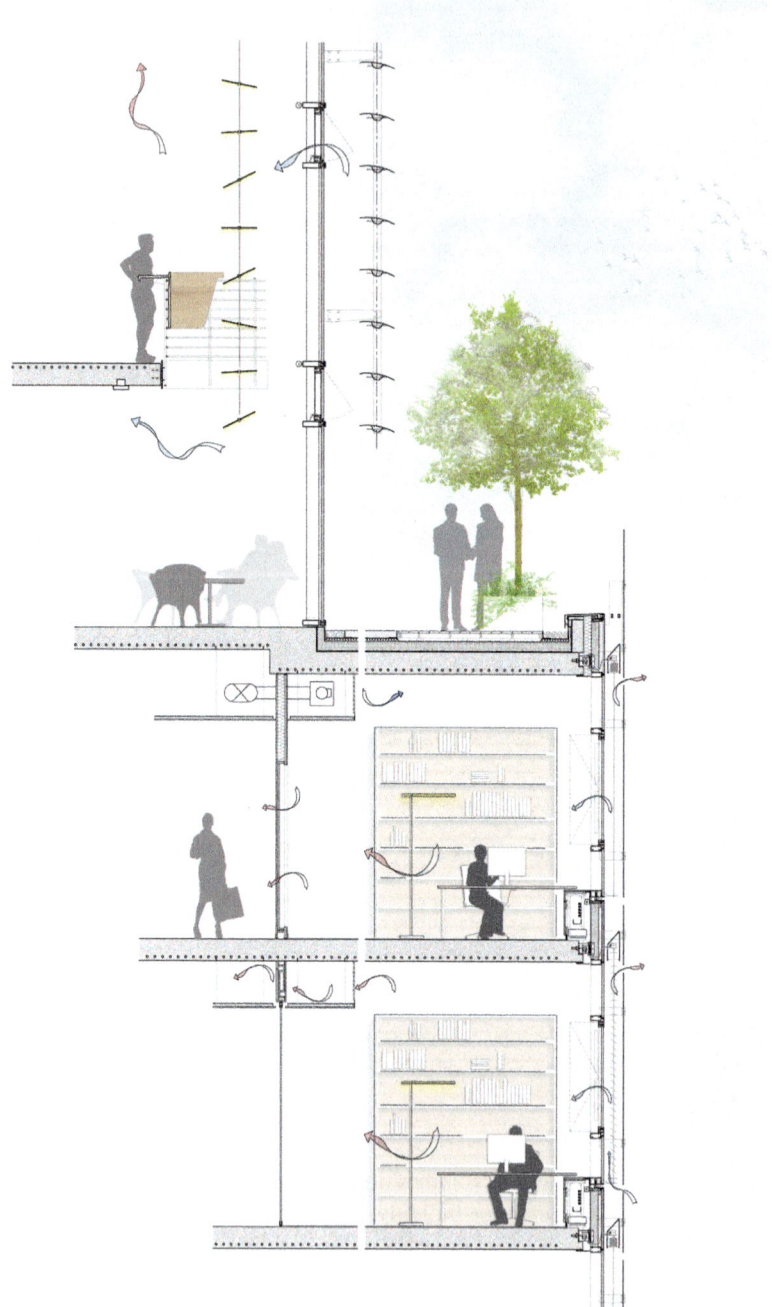

Fig. 1.4 The John & Frances Angelos Law Center atrium facade (*upper floors*) and office/classroom facade (*lower two floors*) wall section. *Image credit* Behnisch Architekten

occupant-aware, daylight-dimming electrical lighting controls. The project is predicted to achieve an Energy Use Intensity (EUI) of 126 kWh/m^2-year (40kBtu/sf-year). If this performance outcome were achieved, the project would meet the energy target of the AIA's 2030 Commitment with a 62.2% carbon emission reduction compared to the Energy Star 50th percentile building.

The concepts implemented in the John & Frances Angelos Law Center are not fundamentally new, but they are not routinely achieved in practice. There is an opportunity to better realize these underlying design approaches in virtually all buildings, not just for special case projects and not simply in the U.S. but globally. Achieving this potential requires addressing a broad set of factors for enabling effective daylighting as a central driver for low-energy, low-carbon buildings. These factors are discussed in the following sections of this chapter.

1.3 Fenestration Design Impacts on Electric Load Shape and Demand Response

Utilities are concerned as much about the timing of electrical energy use as the total use since the power plants and transmission lines are capacity limited. Accordingly, most commercial buildings pay a "demand charge" for peak power use and an "energy charge" for total energy used. Commercial building rate structures are complex, with differential rates for the same unit of energy power consumed during the day vs. night, and summer vs. winter since most utilities experience peak loads during hot summer afternoons. Utilities initially set up special "demand response" programs designed to reduce electric use during the 10–20 peak days of the year. Owners discovered that many of these demand reduction strategies could be used throughout the year. Buildings that can flexibly adjust their electric use in response to price or utility request also have added flexibility in meeting energy and cost performance goals.

Effective daylighting and solar control can play an active role achieving and optimizing the electric load management capabilities of buildings integrated into the expanding "smart grid" of advanced "time-of-day" smart meter infrastructure. This infrastructure enables time-of-use and automatic DR programs which seek to reduce consumer demand for electricity during periods of peak usage or unexpected restrictions in supply. This need will grow in the coming years due to increasing regulatory requirements, increased reliance on time-variable renewable supply and other economic drivers (Fig. 1.5).

Typical commercial building load shapes peak during the afternoon when daylight availability is greatest. Consequently, "daylight autonomous" buildings designed to operate comfortably with minimal electrical lighting during peak demand periods have the potential to significantly reduce loads on the utility grid. However, it is important to simultaneously manage cooling loads due to solar gain. Integrated control of automated solar shading systems, electrical lighting systems,

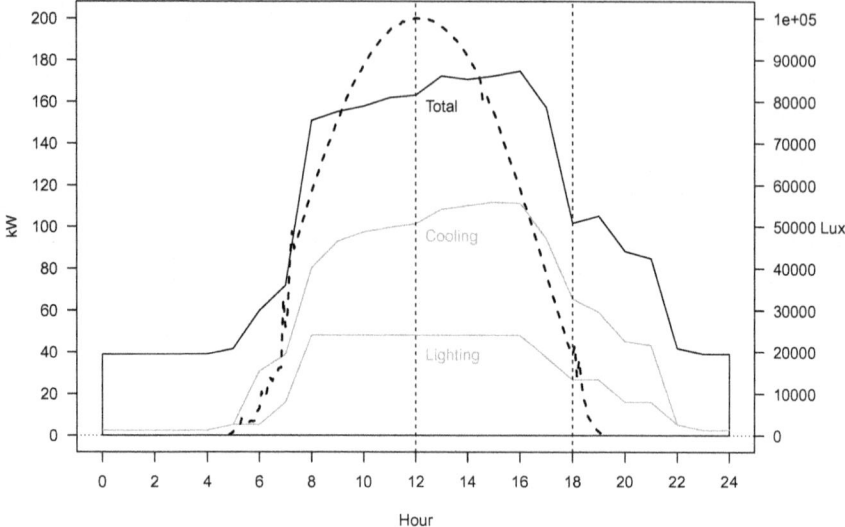

Fig. 1.5 Example load shape of a modern (2004) medium commercial office building located in Los Angeles, CA. Vertical dashed lines from 12:00 to 18:00 indicate the critical "peak" period of statewide demand. *Dashed curve* indicates daylight availability in Lux

operable windows for natural ventilation, and thermal charging/discharging of interior thermal mass has the potential to further increase load-shed and load-shifting capabilities by shifting or reducing cooling loads while maintaining occupant comfort and whole-building energy efficiency.

1.4   Daylighting Impacts on Human Health, Well-Being and Performance

While the optimal management of solar energy is a fundamental consideration for achieving low energy performance objectives, daylighting to support human needs for light in buildings is a far more complex challenge. A growing body of research in the disciplines of photochemistry, photobiology, and human physiology demonstrates that access to daylight and window views have a range of impacts on human health, well-being and performance (Fig. 1.6).

Greater emphasis on the provision of access to window views for all occupants is helping to invert conventional space planning practices for office buildings in the U.S. by placing open-plan offices along the perimeter of the floor plate and locating enclosed cellular office space in the core. For larger buildings, view requirements for the majority of regularly occupied space necessitates a transition from relatively "fat" floor plate buildings with a low surface-to-volume ratio to "thinner" more elongated and complex building forms, with a higher surface-to-volume ratio. Even

Fig. 1.6 Core zone of a large office building lacking access to daylight and views. A homogeneous and steady-state lighting condition is provided by ambient overhead fluorescent lighting. In such environments, lighting provided by electrical sources may be adequate for visual task performance but lack the appropriate spectrum and intensity required to effectively stimulate the circadian system, creating zones of "biological darkness"

in the case of deep floor plate buildings, emerging metrics aimed at quantifying and rating available views, such as the view factor adopted by the LEED rating system (U.S.G.B.C. 2014), serve as an incentive for designers to increase glazed area to achieve required view factors, creating significant technical challenges for managing thermal and visual comfort along the perimeter.

Beyond the needs of the human visual system, the discovery of a novel photoreceptor that mediates the response of the human circadian system to light has led to a growing interest in the "non–visual" effects of light on human health and well being. Much like the ear has dual functions for audition and balance, the human eye has a dual role in detecting light for vision and for adjusting the "circadian clock" which governs most 24-hour behavioral and physiological rhythms (Lockley et al. 2003). Humans possess an internal biological clock that regulates daily patterns of activity following the natural 24-hour light-dark cycle. The suprachiasmatic nuclei (SCN) hosts the circadian clock (or circadian system) responsible for orchestrating the daily timing of biological functions. These functions include sleep/wake, alertness level, mood, hormone suppression/ secretion, and core body temperature. In most people, the period of the SCN is slightly greater than 24 h and relies on patterns of light received at the retina to maintain entrainment with the 24-hour light-dark cycle of the local environment.

In indoor environments, where it is estimated that U.S. adults spend nearly 87% of their lives (Klepeis et al. 2001), lighting is often provided by electrical sources that are adequate for visual task performance, but may lack the appropriate spectrum and intensity required to effectively stimulate the circadian system. In contrast to the visual system, which is maximally sensitive to ($\sim$555 nm) "green" light, the action spectrum of light for the circadian system is shifted towards shorter wavelength ($\sim$460-480 nm) "blue" light (Brainard et al. 2001, Thapan et al. 2001). In addition to spectrum, the intensity of light required to stimulate the circadian system is significantly greater than that required for the visual system for task visibility and must be present for an extended period of time. Therefore, light that may be perceived as adequate for visual tasks may not be effective for circadian stimulus. Further, the time during the day when circadian-effective light is present is important. Bright light in the morning will advance the phase of the circadian system, while bright light in the evening will have a phase-delaying effect. Over time, lack of sufficient exposure of the retina to bright, circadian-effective light can disrupt the circadian system, which can in turn lead to poor sleep, reduced performance, and increased risk of a range of health maladies including diabetes, obesity, cardiovascular disease and cancer (Zelinski et al. 2014). While the underlying science is convincing that these are important effects, the specific cause/effect relationships, the overall impacts on occupants and the appropriate design responses are still evolving (Fig. 1.7).

Until recently, conventional light sources could not readily control the variable spectrum and intensity needed to address these biological needs. The electrical lighting industry is now beginning to promote Solid State Lighting (SSL) technologies as a vehicle to more easily change the spectral content and intensity of light than with gas discharge sources and thus should be more effective for maintaining circadian entrainment in buildings. It is now technically possible for an RGB-based LED to match any desired equivalent daylight color and intensity with the right sensors and controls, and SSL task lighting can be used to produce a circadian-effective light stimulus over a small area most relevant to an occupant's required vertical dose at the eye. However, these approaches require substantial cost and effort, and providing the vertical dose at the eye places significant restriction on occupant mobility.

Daylight is an attractive alternative to electrical lighting for maintaining human circadian entrainment indoors due to its spectrum, intensity, general availability, and potential to be introduced into spaces via windows and skylights. Enabling designers to achieve and optimize "circadian effective" daylighting strategies will require a new set of performance objectives, measurement techniques, assessment tools, and design strategies to ensure the appropriate spectrum, timing, intensity, and duration of light is delivered to maintain healthy circadian entrainment. Figure 1.8 presents an example application (described in greater detail in Sect. 2.4.3) of an emerging simulation-based approach to assess and summarize the circadian effectiveness of eye-level daylight exposures over an annual period within a space.

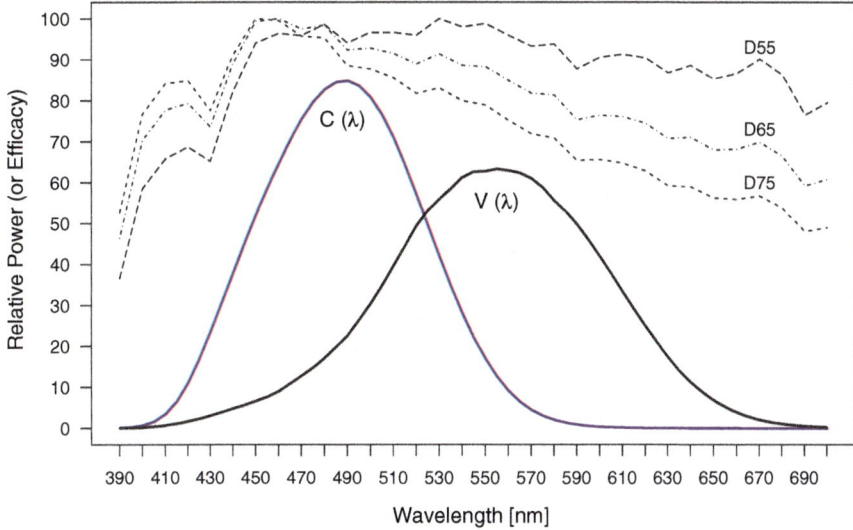

Fig. 1.7 Comparison of spectral response of the visual (photopic) system (V-Lambda) and the circadian system (C-Lambda) to the relative spectral power distributions of three CIE daylight illuminants: (D55) sunlight, (D65) overcast sky, and (D75) north sky daylight. *Note* Both *response curves* are scaled to have equal area under the curves

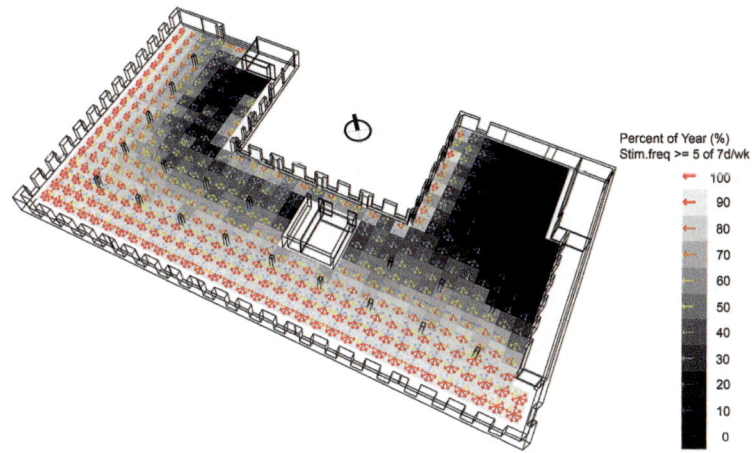

Fig. 1.8 Perspective view of building floor plate located in Downtown Los Angeles showing annual result for the percentage of analysis hours during the circadian resetting period (7:00–10:00 AM) where a minimum stimulus frequency of 71% (5 of 7 days/week) was achieved

Spatial-based exposure results can be used to identify, quantify and visually examine building zones where long-term occupancy may lead to disruption of the circadian system in the absence of supplemental electrical lighting capable of effective circadian stimulus, or other daytime exposure. While theoretical knowledge and scientific findings are sufficient to begin to propose metrics and procedures to classify indoor daylit spaces in terms of anticipated circadian effectiveness, this remains an emergent and active research area.

1.5 Design for the Next Century

The use of daylight to reduce energy consumption and enhance Indoor Environmental Quality (IEQ) is one of the most common claims made for commercial office buildings promoted as "sustainable", "energy efficient," "green," or "high performance." Claims of successful daylighting are often based on the use of large areas of high Visible Light Transmittance (VLT) facade glazing, photo-controlled electrical lighting systems, and results from lighting simulations performed during design that demonstrate compliance with green building rating system criteria (e.g. USGBC LEED Daylight and View EQ credits). Design decisions are often based on the assumption that making the building envelope as transparent as technically possible will lead to an increase in the amount of interior daylight available, leading to greater levels of occupant satisfaction and visual connection to the outdoors. But it then becomes a tremendous technical challenge to provide thermal and visual comfort immediately adjacent to a floor-to-ceiling glazing design and to address HVAC loads from the glazing.

Buildings are rarely studied in use to examine if "transparent" facades achieve design intent for overall energy and occupant comfort and performance, even if they achieve the goals of high visual transparency or daylight transmission and meet minimum code requirements. When conducted, Post Occupancy Evaluation (POE) studies often demonstrate that "unshaded", highly glazed facades produce indoor environmental conditions which are often visually and thermally uncomfortable (or, at times, intolerable) for occupants, leading to formal and informal modifications to the facade that can significantly limit anticipated energy reduction and IEQ benefits. Unshaded highly glazed facades are also likely to have high heating and cooling loads, depending further on climate and orientation. The result may be a transparent facade "design" that is made largely semi-transparent or even opaque "in operation" by occupants to reduce discomfort. Figure 1.9 shows the operational outcome for a "transparent" southwest-facing facade located in San Jose, California, a climate with predominantly clear skies. The southwest facade was photographed informally over more than four years, where interior shades were observed to occlude the majority of facade glazing, and remain static in place for months, or in some cases years (Fig. 1.10).

Fig. 1.9 Common outcome of a "transparent" facade made largely opaque by its occupants, limiting daylight transmission and views to the outdoors. South-west facing facade of commercial office building in San Jose, California, where sky conditions are typically clear. *Image credit* Prof. Charles Benton

Contemporary daylighting design practices, which favor highly glazed "transparent" facades, emerge in part from the relatively cool, heating-dominated climates of northern Europe, which have predominantly overcast skies and low demand for air conditioning. Due to a lack of effective shading, even in moderate U.S. climates with significant hours of sun, such as San Jose, CA or Los Angeles, CA, facade solar heat gains lead to significant cooling loads that are conventionally offset by the use of air conditioning. In contrast to the location of most existing "transparent" architecture, the majority of future growth in the 21st Century will be in much warmer climates. The export of contemporary "transparent" architectural design features to these locations, without any compensation for climate, will have significantly greater adverse effects on energy (and carbon) outcomes due to the level of air-conditioning needed to make such buildings operable, combined with the generally greater carbon intensity of the electricity supply in many regions. Using the simple metric of Cooling Degree Days (CDD) as an indicator of annual cooling demand, Fig. 1.11 compares the cooling demand in the U.S. cities San Jose (the location of the example shown in Figs. 1.9 and 1.10) and Los Angeles (in light blue) with the regions that are anticipated to experience the majority of urban growth in the 21st century (dark blue). The current population (in millions) of each region is shown in parenthesis.

Alternatively, integrating a high-transparency glazed envelope that provides daylight with a design that incorporates effective solar control to enable passive and

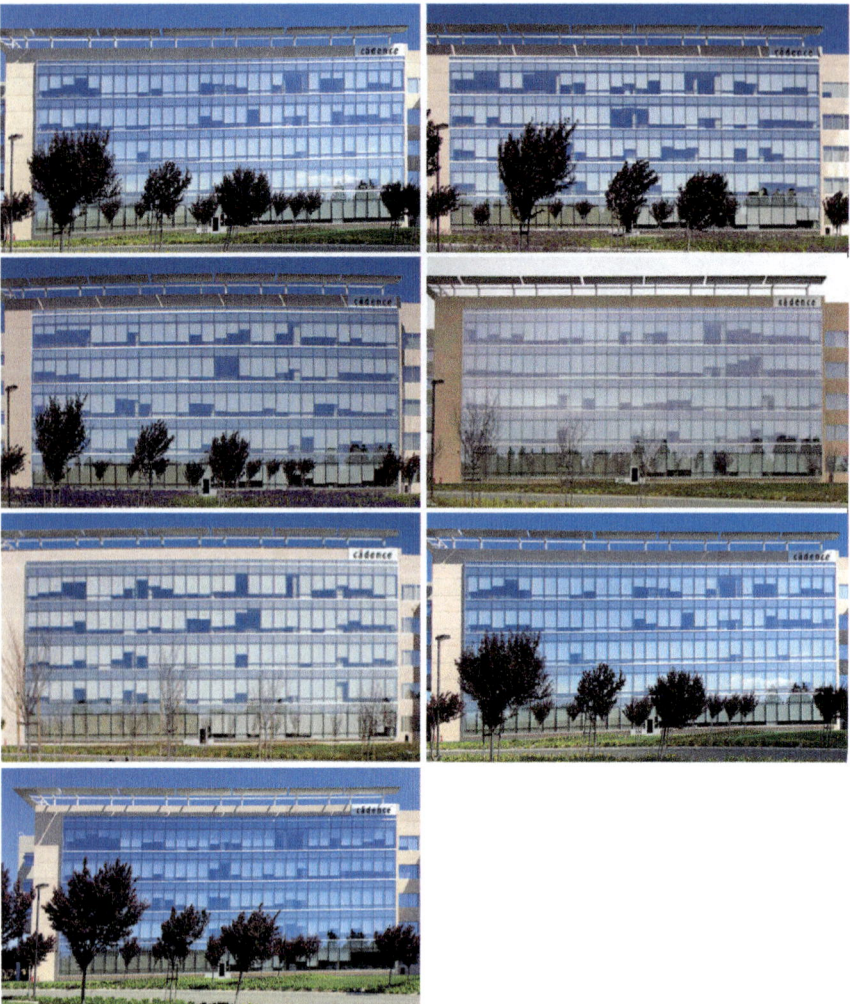

Fig. 1.10 Example operational outcome for a "transparent" facade located in San Jose, California, a climate with predominantly clear skies. Date of each image, clockwise from top left, August 2009, December 2009, January 2011, October 2012, April 2014, August 2011, August 2010. *Image credit* Prof. Charles Benton

low-energy cooling can provide not only the daylighting benefits but the co-benefit of greater operational reliability (i.e. "passive survivability") during potential interruptions to the electricity grid as well as greater potential for demand response to manage time dependent electric load. For example, consider the floor plate of the 10-story, 22,500 m^2 office tower in Canberra, Australia, which is sidelit on three sides by a floor-to-ceiling glazed facade curtainwall (spectrally-selective low-e facade glazing (VLT 62%, SHGC 0.28, u-value 1.64 W/m^{2}K)). Facade glazing is

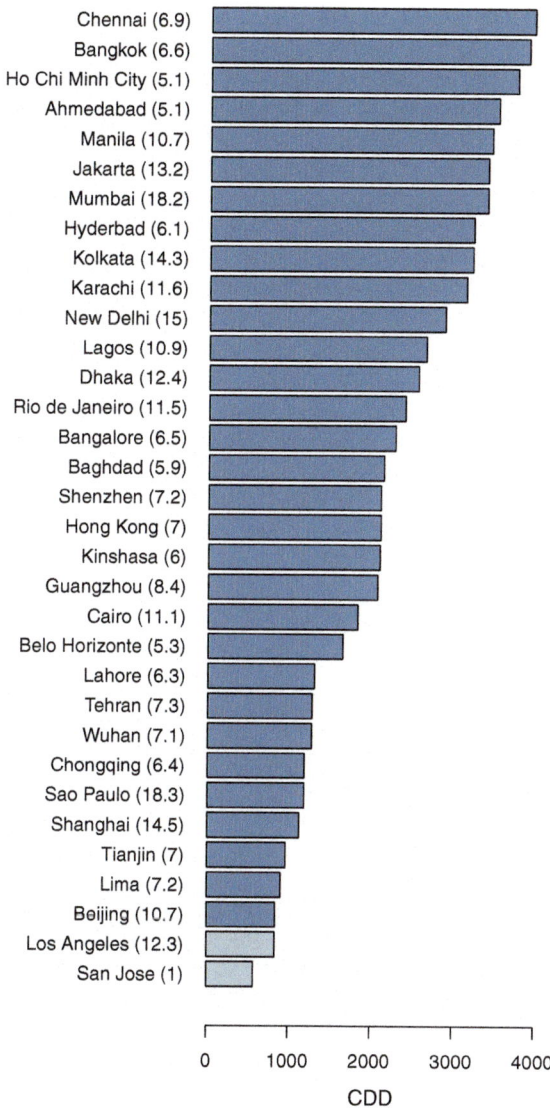

Fig. 1.11 Comparison of annual cooling demand (reported in terms of Cooling Degree Days (CDD)) for the U.S. cities San Jose and Los Angeles with the regions anticipated to experience the majority of urban growth in the 21st century. Figure after "Crank it Up," published August 18, by the New York Times (NYT 2012). Data shown in dark blue originally from Michael Sivak, University of Michigan (http://www.nytimes.com/interactive/2012/08/19/sunday-review/19rosenthal-ch-int. html?action=click&contentCollection=Sunday%20Review&module=RelatedCoverage&region= EndOfArticle&pgtype=article)

Fig. 1.12 Exterior solar control screen of the NewActon Nishi office building in Canberra Australia. The exterior screen is designed to limit peak solar radiation to 60 W/m^2, enabling the application of low-energy environmental control strategies (natural ventilation, automated night flush cooling of exposed concrete thermal mass and ventilation via an underfloor air distribution system). *Image credit* Carl Drury

shaded by external fixed horizontal wood louver screen (Fig. 1.12) engineered to provide sufficient solar control to enable the application of passive and low-energy cooling strategies as an alternative to forced-air HVAC (Fig. 1.13). The blocking angle for the louvers was calculated to limit peak solar gain to 60 W/m^2. This was done so that a high efficiency/low temperature under-floor air system, paired with natural ventilation and night-flush cooling could be utilized while still meeting peak

Fig. 1.13 View along integrated access way showing open state of automated *upper windows.* *Image credit* Carl Drury

cooling demand. Analysis was done in proprietary engineering software to study various external shading strategies (horizontal and vertical) and glazing combinations using Canberra climate data. Additional glare control is provided to occupants by manually operated interior roller shades (VLT 6-9%).

Excluding renewable energy generated on site, in 2015 the building resulted in a measured (and publicly disclosed via the Australian Government's Commercial Building Disclosure (CBD) program) annual energy consumption of 1661,000 kWh (74 kWh/m^2-year), and an annual carbon emission intensity of 46 kgCO^2-e/m^2, making it one of the most resource efficient commercial buildings in Australia.[4]

1.6 Challenges of Time and Scale

The challenge of effective daylighting lies not only in achieving low energy outcomes that simultaneously support end-user psychological and physiological needs for light, but in doing so rapidly and at scale to avoid locking-in the current inefficiency of Business As Usual (BAU) practices for the next 50–100 years. While the topic of daylighting in architectural publications often focuses on unique building types or newly constructed high-budget projects, addressing the problem of scale requires

---

[4]https://cbdportal.cbd.gov.au/Download/ShowPdf?id=B1800-2015-1.

practices that can be replicated broadly across a range of project types, regions and budgets. Scalability is necessary to meet carbon mitigation goals and important for ensuring equitable access to indoor environments that support high levels of health and well-being for occupants. Consequently, designers are faced not only with developing innovative, new prototypes and practical strategies for retrofits to the existing building stock, but in demonstrating the effectiveness of performance outcomes to stimulate market adoption. To illustrate the magnitude of the challenge using an example from California, the existing California building stock must become 40% more energy efficient by 2030 to achieve statewide greenhouse gas emission reduction targets (e.g. Executive Order S-3-05). Taking into consideration only the commercial building stock existing today (totaling ~ 465 million square meters (~ 5 billion square feet) as of the most recent survey, completed in 2006), achieving this target would require deep-energy retrofits to 36 million square meters of commercial buildings each year beginning now (i.e. 2017). Promoting widespread energy efficient design practices in the 21st century requires the creation of a body of evidence demonstrating that practices lead not only to reliable performance and more efficient resource use, but also to indoor environmental conditions that are preferred by occupants over conventional sealed and mechanically controlled environments.

1.7 Defining Effective Daylighting

One of the central barriers to effective daylighting is that daylighting *performance* is often defined differently by different stakeholders, leading to a fragmented approach to performance assessment in the design and operational life-cycle of buildings. For example, a mechanical engineer may define performance in terms of achieving low Energy Use Intensity (EUI) or Zero Net Energy (ZNE) whole-building performance. Alternatively, an architect may define performance in terms of the aesthetic qualities of daylight distribution in the space or the perceived level of visual transparency of the building facade. The client may define performance based on whether or not the project complies with the requirements of green building certification criteria for daylight sufficiency and views. Finally, building occupants may judge daylighting performance based on their perception of daylight sufficiency, visual comfort, and available views or the level of controllability provided by the design to adjust and adapt to dynamic environmental conditions at their workspace. Thus, daylighting performance encompasses a range of factors that, if considered in isolation, can lead to misleading conclusions. A space that "maximizes" daylight transmission to reduce electrical lighting energy consumption but results in visual discomfort may lead to constant use of interior shading devices as well as ad hoc and formal modifications to the facade (or workstations), which significantly changes the design intent.

At a fundamental level, effective daylighting can be defined as "building designs that deliver on performance goals." This will require new or enhanced practices that simultaneously embrace three central elements: (1) daylighting design objectives that support a low energy concept, (2) design strategies that routinely meet end-user needs for daylight access, views, visual/thermal comfort, and personal environmental control and, (3) feedback mechanisms that are applied during the design, delivery and operational stages of the project to align performance in use with design intent.

It is common today to hear about the processes of "integrated design," or "multi-disciplinary collaboration", which have become widely promoted for improving whole-building energy efficiency through greater collaboration of various project team members and system integration during the design stage. Effective daylighting expands the concept in two significant ways. First, effective daylighting incorporates validation and learning during the delivery and operational stages of projects, with the goal of creating a body of empirical evidence from the field that can be leveraged by design disciplines to inform future projects and practices. The Architecture, Engineering and Construction (AEC) industry is extremely risk averse, and slow to adopt promising technologies and design strategies without proof from real buildings in use demonstrating both energy performance and high levels of end-user acceptance. Therefore, trustworthy feedback in the form of measured data that supports validation and learning are critical for differentiating innovative design practices, identifying technologies that work, correcting failures, improving market adoption, and broadly disseminating knowledge to improve design practices and technology performance specifications. Second, where integrated design is traditionally focused on a narrow objective of energy optimization, effective daylighting includes multiple human-factors performance goals as well as novel feedback mechanisms to position end-users as a central indicator of project performance (see Chaps. 2 and 6). Finally, accelerating the flow of knowledge and experience across this "design-operations" feedback loop should also pressure and encourage the building industry to innovate more rapidly and successfully to deliver new integrated facade technologies and systems that are more reliable and lower cost, thus making it easier for teams to achieve their design goals.

1.8 An Agenda for Effective Daylighting

Enabling broad application of effective daylighting requires an agenda for addressing factors that currently limit optimal utilization in contemporary design practices, project delivery and performance in use. The following sections frame an agenda within the context of three central transformations:

1. From compliance-based to performance-based design.
2. From static and unresponsive to context-aware and adaptive systems.
3. From theory to validation, feedback and learning.

1.8.1 From Compliance-Based to Performance-Based Design

Daylighting has been seen as an energy efficiency strategy since the oil embargo of the 1970s and its relative importance has evolved over time. There are a growing number of new incentives for the use of daylight as a strategy for electrical lighting energy reduction and enhanced IEQ. These include green building rating systems (e.g. LEED), standards for the design of energy efficient buildings (e.g. ANSI/ASHRAE/USGBC/IES Standards 90.1-2013 and 189.1-2014) energy code lighting power adjustments for photocontrolled electrical lighting (CA Title 24 Building Energy Efficiency Standards for Non-Residential Buildings) and emerging standards focused on occupant health and well being (e.g. the International WELL Building Institute's WELL Building Standard). However, because these are fragmented (and often conflicting) objectives, projects designed to achieve various compliance criteria often fail to integrate daylighting goals within a whole-building energy strategy or make optimal use of the daylighting potential of the local climate to serve the diverse array of end-user needs for daylight. This is largely due to the fact that many of the design decisions occur after design development, for code-compliance purposes or to obtain green building certification rather than during the early stages of design, where the largest impacts on project performance are established. It is critical to have well-defined performance goals at the start of design and for performance evaluation to be integrated into the planning and schematic phases of design, where feedback from analysis can inform design decision-making to improve the environmental quality and energy performance of the project.

The emergence of whole-building low energy and ZNE building performance requirements combined with a growing array of human-factors objectives are driving a reversal of the conventional process of project design and performance analysis. Rather than using analysis to confirm that a predetermined design complies with various code and green building criteria, practitioners and researchers are now exploring how iterative simulation-based analysis can be used in early stages of design to rapidly identify optimal performance outcomes among multiple competing design options, and to examine trade-offs between conflicting performance goals. Performance-based design promotes the exploration of building forms, fenestration systems and controls that are "tuned" to the specific climatic, programmatic and contextual conditions of each project to optimize the use of climate for both IEQ and whole-building energy performance objectives.

At the most fundamental level, a performance-based design process is defined by a feedback mechanism utilizing analysis tools to relate prospective design strategies with measureable project outcomes (Fig. 1.14). By examining how design decisions impact project performance, particularly in early stages of design, knowledge can be generated and fed back to inform decision-making with the objective of improving the performance of future design iterations. Whole-building energy performance specifications, building energy benchmarking and public disclosure requirements, along with outcome based codes and energy-performance-based

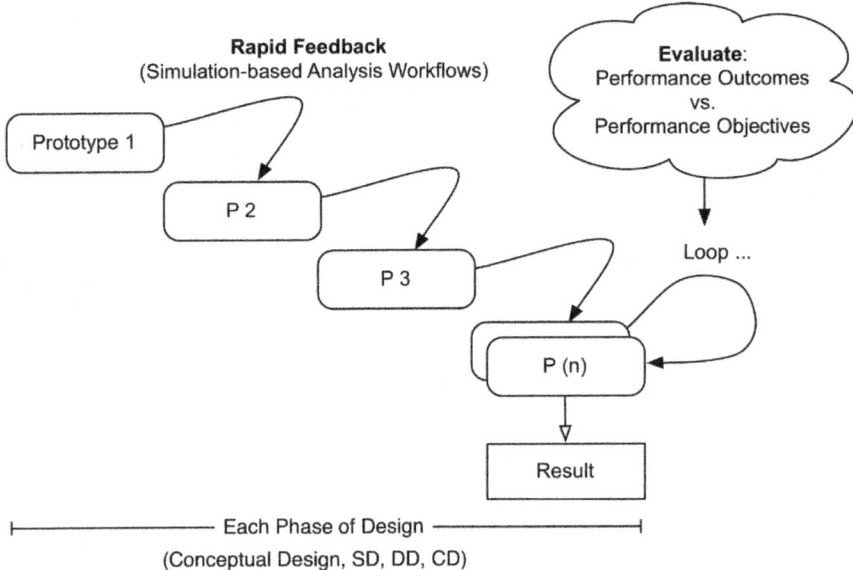

Fig. 1.14 Fundamental organization of an iterative, performance-based design process

procurement[5] adds additional incentive for design firms to seek mechanisms for reliable, early-stage performance feedback.

Implementing performance-based design in practice requires the development of new simulation-based workflows (Fig. 1.15) combining 3D authoring software with energy and lighting simulation engines, accurate thermal and optical data on all the design elements, as well as optimization and visualization tools that are capable of providing rapid and reliable analysis feedback at the pace of the design decision-making process. As projects targeting low-energy goals often implement passive environmental control strategies (e.g. solar control, natural ventilation, thermally charged/discharged mass, daylighting), which must be carefully designed in response to local climate and context, simulation tools must be capable of reliably modeling the effects of the local climate and context as well as the behavior of passive systems.

There is no single optimization tool available to translate project objectives and constraints into a holistic design outcome. Nor is there consensus for how to best manage trade-offs among various performance objectives, or how to assign relative weighting to performance metrics based on their perceived importance among various project stakeholders (e.g. design team, project manager, or end users). In the real world, one needs someone to sort the global problem into chunks that can be

---

[5]A procurement process where project teams are selected based on the predicted performance of a proposed design, and contractually obligated to deliver a project that performs within the range predicted.

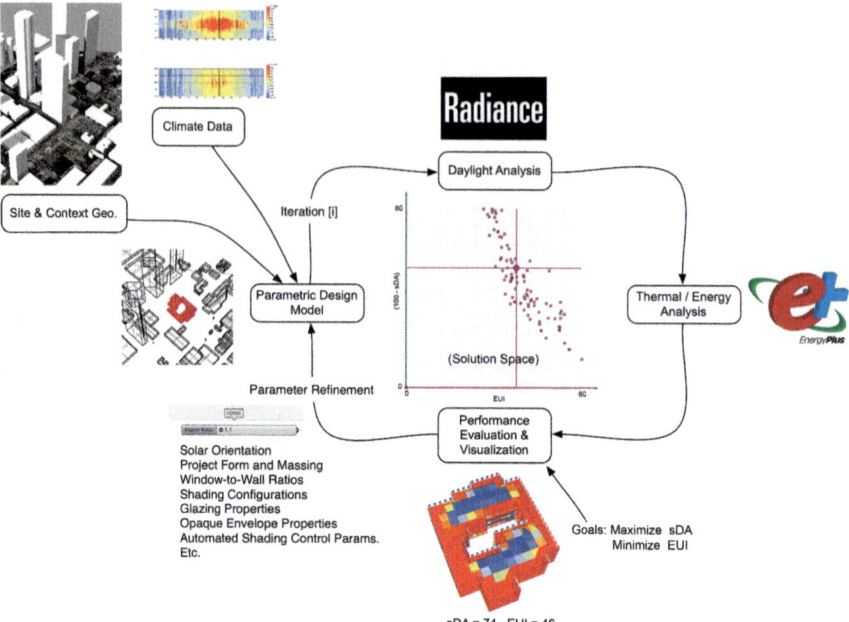

Climate Data

Radiance

Daylight Analysis

Site & Context Geo.

Iteration [i]

Parametric Design Model

(Solution Space)

Thermal / Energy Analysis

EnergyPlus

Parameter Refinement

Solar Orientation
Project Form and Massing
Window-to-Wall Ratios
Shading Configurations
Glazing Properties
Opaque Envelope Properties
Automated Shading Control Params.
Etc.

Performance Evaluation & Visualization

Goals: Maximize sDA
Minimize EUI

sDA = 74 EUI = 46

Fig. 1.15 Example performance-based design framework linking highly optimized lighting and thermal analysis with 3D parametric modeling, visual scripting and optimization tools for rapid prototype development

analyzed and optimized using available tools and guidance and then recombine those chunks into a coherent overall package. This is an ongoing and evolving process—it needs to be initiated at one level of detail in early design/schematics and then continued later (perhaps on multiple occasions) through DD, CD, VE and even late in construction. Figure 1.15, (discussed further in Chap. 4), presents an example implementation of a performance-based design framework linking highly optimized lighting (Radiance) and thermal analysis (EnergyPlus) simulation engines with 3D parametric modeling, visual scripting and optimization tools for rapid prototype development in early-stage design.

This book addresses the performance of dynamic daylit spaces from a broad perspective that includes assessment of occupant behavior, occupant subjective assessment of daylight sufficiency, view, visual and thermal comfort within a whole-building energy concept. Occupant behavior and human-factors metrics are discussed within a framework of design workflows, visualization techniques and novel "in-situ" Post Occupancy Evaluation (POE) methods capable strengthening the feedback loop between design intent and performance in use. A critical analysis of energy and human-factors performance metrics is presented in Chap. 2. Chapter 4 provides a discussion of how metrics, analysis tools and workflows are being applied within emerging performance-based design frameworks. As of the publication date of this book, all of these design processes and approaches, as well as the underlying tools, are in a continuous state of active development and refinement, promising new, enhanced options in the future.

1.8.2 *From Static and Unresponsive to Context-Aware and Adaptive Systems*

Contemporary approaches to daylighting design often implement static facade systems, which are incapable of responding to daily or seasonal changes in sun and sky conditions or effectively managing between the dynamic range of outdoor solar and lighting conditions and the range indoors that occupants require (or prefer). While static facade systems may serve as a practical option for some lighting and HVAC energy reduction efforts, the resulting indoor environmental conditions are often unacceptable to occupants for significant periods of time. Furthermore, while static solutions may be "adequate" for small fenestration areas that just meet compliance codes, they fall short of highly glazed designs that typify many attempts to extend daylight impacts in low energy buildings. As a result, static facades that "optimize daylight" through maximizing physical transparency often lead to retrofits and occupant modifications over the project life cycle to address glare and solar overheating which, in turn, serve to greatly reduce the anticipated energy savings and IEQ benefits. Alternatively, static facades that incorporate extensive fixed shading, small window apertures, and glazing technologies to reduce visual transparency fail to achieve energy (e.g. ASHRAE 90.1) or IEQ (e.g. LEED EQ) objectives. As the architecture, engineering and construction industries shift

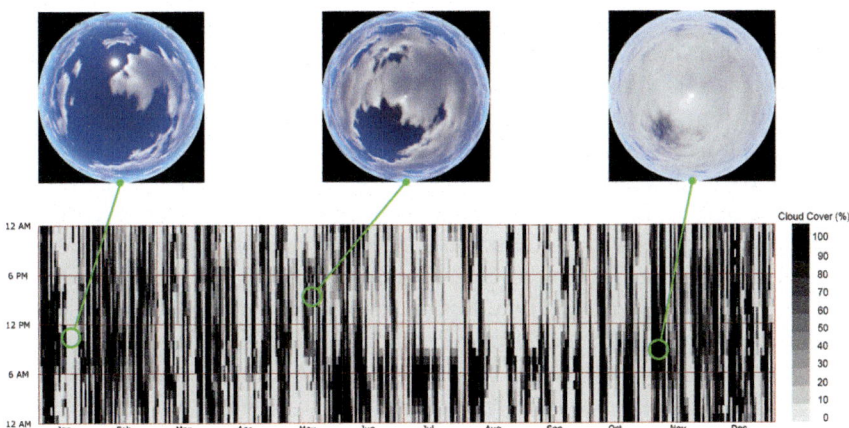

Fig. 1.16 Annual hourly cloud cover (0–100%) for San Francisco, CA

towards pursuing low and ZNE design strategies as standard practice, it is anticipated that design teams will increasingly explore the integration of dynamic,
environmentally responsive facade technologies to achieve greater levels of
building performance and occupant needs.

Dynamically responsive facades are needed due to the fundamentally dynamic
nature of the sun and sky. Figure 1.16 shows the annual hourly cloud cover and
example sky conditions for a location in San Francisco, CA. In concept, dynamic
facade systems are capable of continually adjusting the envelope features to seek
the optimal balance between energy and human-factors objectives for any given sky
condition. However, for dynamic facade strategies to perform effectively over their
life cycle requires the development of systems that are capable of modulating
exterior conditions to deliver the indoor environmental conditions desired by
building occupants.

Active use of the building envelope (e.g. solar control, daylighting, natural
ventilation, and charging/discharging thermal mass, energy harvesting) paired with
controllable lighting and HVAC systems is a complex design challenge. However,
driven in part by typical building codes, application of building technology often
focuses on the efficiency of individual components rather than consideration of the
overall performance of multiple components working as a system. This fragmented
approach needs to shift to an integrated, context-aware dynamic perspective that
addresses the facade as a system that is responsive to "performance needs" at three
different levels: (1) comfort and task performance needs of the occupants; (2) energy and economic needs of the building operator; and (3) the local or regional
needs of the utility grid.

While significant effort has been placed on "integrated design" practices that
seek to achieve greater levels of system integration during the design stage, the
operational performance of integrated systems in the occupied building is limited
by a number of barriers. These include (1) the lack of interoperability between
various technologies, (2) challenges in deploying and maintaining large sensor
arrays (e.g. unit cost, commissioning, calibration), (3) lack of detailed, granular,
contextual data to drive effective real-time operation, (4) poor or non-existent
mechanisms for fault detection and diagnostics, (5) lack of occupant feedback to
validate controls assumptions or make adjustments, and (6) lack of holistic controls
optimization frameworks (due in large part to #5). From a process point of view,
design concepts may not be adequately conveyed to and implemented by the
construction team, and the hand off to facility managers and occupants is often
incomplete and imperfect. Improvements and innovations in the technology systems are further limited by, (1) the lack of frameworks for systems to gather and
interpret performance data and learn over time, and (2) the lack of a mechanism to
store and share knowledge across projects and design team members.

The result of these limitations has been failures in building performance and a
resultant aversion among building designers and contractors to adopt complex but
promising technologies in favor of "simple" control strategies based on the cautionary view that "simple is always better." Entirely passive, fixed solutions seem
unlikely to properly address the wide range in climate and user needs in most

buildings. Asking occupants to become de facto facility managers and adjust light levels, blinds, thermostats etc. seems equally unlikely in the majority of buildings. However, fully automated systems risk alienating occupants when they fail to deliver desired comfort conditions. The real world perspective also suggests that occupants may adjust building features for comfort, but will not reliably manage energy performance objectives. We suggest it is time to challenge the common knowledge that "complex controls will never work" and that hybrid models cannot be adapted to support local occupant needs.

The sensors and controls industry globally is now in the midst of a revolutionary change driven in part by the rapid advance of the "Internet of Things" (IoT) movement. The Internet of Things is the network of physical objects—devices, vehicles, appliances and other items embedded with electronics, and sensors, and linked by software-based network connectivity—that enables these objects to collect and exchange data, and then act based on that data. IoT is based on four critical elements: (1) low cost, distributed powerful sensors and embedded computing, (2) wireless communications; (3) cloud based data storage and computation, and (4) shared interoperable protocols. Much of this technology and infrastructure was created and driven initially by the smart phone industry, but is rapidly gaining traction in numerous other business realms including the building industry where the LED revolution in the lighting community is leading the way. It will likely be further accelerated by massive RD&D investments underway to develop autonomous vehicles where distributed sensing and controls-well beyond the needs of a dynamic building envelope-will need to be developed and perfected and manufactured in volume.

Figure 1.17 presents a conceptual framework for the design of IoT-enabled Perimeter Systems (IoTePS). The IoT movement can be leveraged within the building design domain to develop context-aware, interoperable building components that work to optimize the comfort and resource efficiency of buildings throughout the project operational life-cycle. The IoTePS framework is conceived as a vehicle to explore how the ubiquity of sensing, real-time data and computation will transform existing approaches towards building facade and perimeter zone technologies and the performance roles those technologies are asked to play in buildings. Of specific interest is the transformation of the building facade from a sealed and static element to a dynamic filter, operating in real time to manage a range of grid-level, building-level, and occupant-level performance goals. Charting the functional potential of dynamic behavior, informed through detailed real-time and historic sensor and occupant feedback data, will in turn serve as a basis to explore and develop new specific architectural fenestration strategies, (both technologies and design approaches), to best meet this potential.

The current challenge is to create integrated systems that are capable of delivering acceptable (or preferred) environmental conditions to occupants over an annual range of environmental conditions while simultaneously contributing to

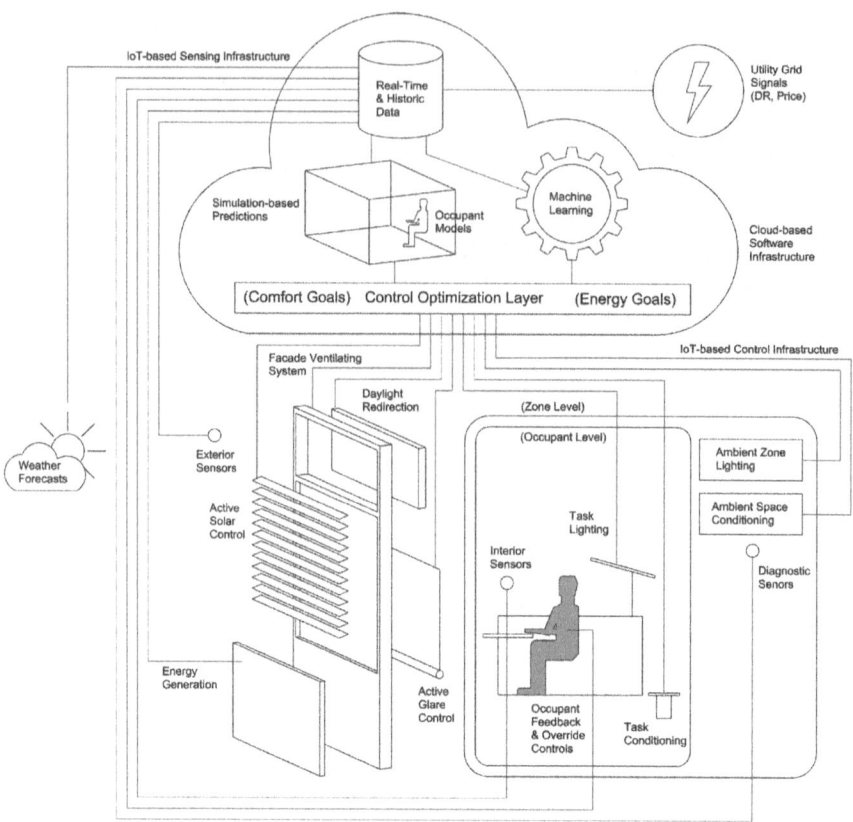

Fig. 1.17 Conceptual organization of an interconnected, human-in-the-loop facade and perimeter zone system

whole-building energy performance goals. Effective operation of automated systems requires that external and internal environmental parameters be accurately sensed, that control assumptions are validated against data-driven models of occupant behavior and subjective preferences in order to ensure long-term user acceptance, and the hardware and software solutions can be fabricated, installed and calibrated on time and on budget. As part of these solutions, appropriate user interface technologies are needed to easily integrate occupants as a mechanism for user-overrides. Achieving these objectives requires going beyond the physical integration of components in construction. The book presents emerging and novel strategies to shift from closed-loop systems and ad hoc control assumptions to context aware, humans-in-the-loop systems by leveraging the growing availability of low-cost sensing and internet-connected devices to develop interactive, interconnected systems capable of learning and adapting to changing contextual and environmental factors (e.g. Fig. 1.17). These topics are outlined and discussed in Chap. 3.

1.8.3   From Theory to Feedback, Validation and Learning

As designers seek to integrate daylighting within an efficient whole-building energy strategy, how best to manage trade-offs between objectives such as envelope thermal performance, lighting and HVAC energy demand with human factors such as visual and thermal comfort, daylight availability, visual connection to the outdoors, and personal control requires an approach informed at a fundamental level by empirical knowledge of end-user needs and behaviors. Even in the most sophisticated simulation tools and workflows, the presence and environmental preferences of occupant are often represented by crude, static and universally applied assumptions. In practice, crude application of human factors data limits the energy and carbon reduction potential of energy efficiency measures, and can lead to operational challenges and discrepancies between anticipated and measured energy consumption. Although it is unrealistic to assume that the preferences and behaviors of a specific population of building occupants can be routinely predicted with a high degree of accuracy, (particularly prior to construction of the project), it is important for designers to be aware of the large array of human-factors assumptions embedded in software-based design tools and understand the impacts these assumptions may have on anticipated performance outcomes.

Existing lighting design metrics are based on a legacy of controlled human-factors laboratory experiments yielding universal design guidance. This guidance, originally intended for electrical lighting design applications, is not well suited to the design of daylit spaces. In contrast to the static and spatially-homogeneous conditions produced from electrical lighting, daylit spaces respond dynamically to hourly, daily and seasonal changes in sun and sky conditions, and generally produce higher luminances and luminance contrasts throughout the space due to the greater intensity of light from the sun and sky as well as the location of fenestration in the occupants' vertical field of view (Fig. 1.18).

Although there is growing consensus for the importance of daylight and views in commercial buildings, there is less consensus for how performance objectives such as daylight sufficiency, visual comfort, and view should be defined, measured, relatively valued, and how results should be interpreted over an annual basis to assess success or failure. Consequently, designers are unable to reliably assess end-user outcomes during design, or optimize a design to balance energy objectives with occupant comfort. How building occupants accept, adapt to, and modify

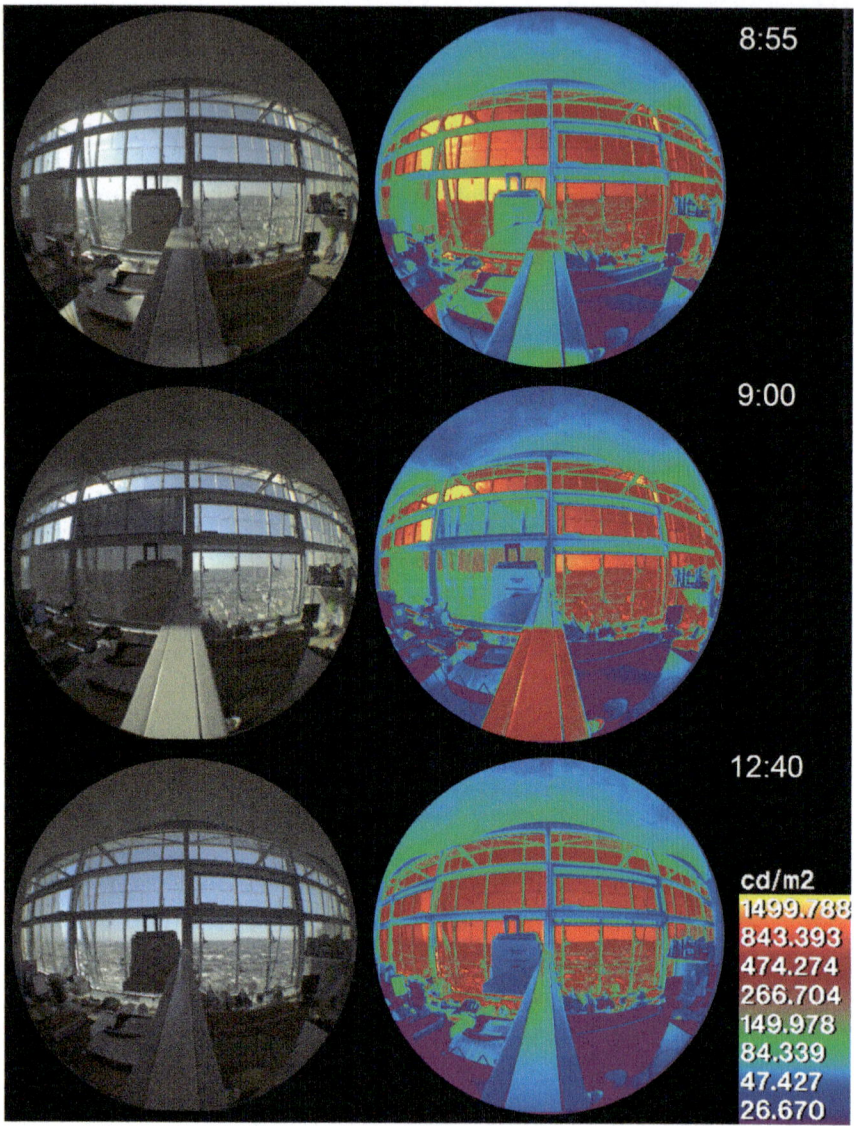

Fig. 1.18 Dynamic daylighting and glare conditions observed using "in-situ" High Dynamic Range (HDR) camera monitoring equipment in a southeast facing perimeter zone of an office building located in San Francisco, CA on August 25 under predominantly clear sky conditions. Right column display the luminance of each pixel using a falsecolor tone-mapping (logarithmic). *Horizontal rows* indicate luminance conditions (and times) before and after occupant adjustment to facade shading devices

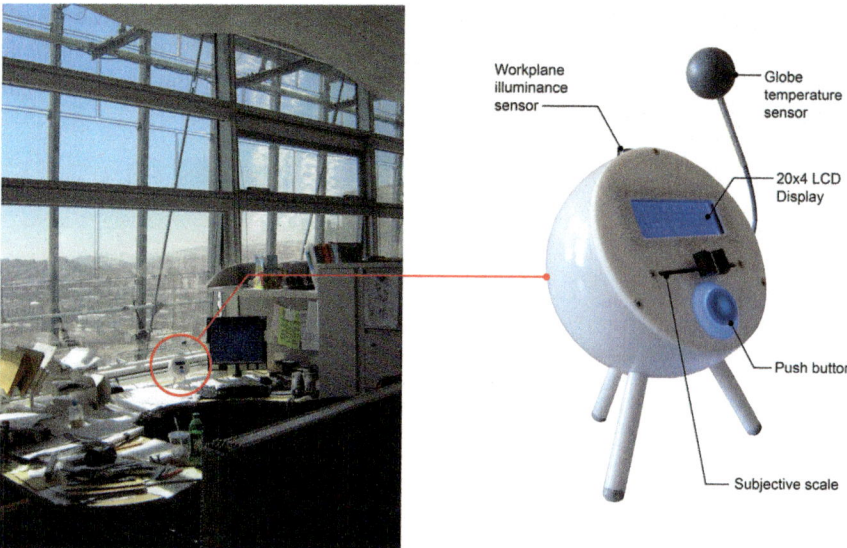

Fig. 1.19 Example "in-situ" method of human-factors data collection in buildings in use

dynamic daylighting environments over time is a difficult phenomena to examine in a controlled laboratory setting, and leads to the need for "nomadic" field research methods and continuous commissioning technologies to build a body of evidence for appropriate human-factors design parameters.

Closer consideration of occupant experience in buildings is integral to meeting the need for resource-efficient and climate-resilient buildings. Rather than passive recipients of indoor environmental conditions, occupants represent a rich multi-sensory source of information on environmental performance with the potential to serve as vital resource to better understand and respond to the complex relationship between the built environment and its inhabitants. This book discusses the application of emerging "in-situ" methods (e.g. Fig. 1.19) to collect detailed feedback data pairing physical measurements from the indoor environment with subjective feedback from building occupants in real time. Enabling real-time feedback from building occupants paired with granular physical measurements has the potential to significantly advance the ability of design teams, commissioning agents, and building operators to assess, benchmark, and learn from innovative projects and to continually optimize efficiency goals with occupant comfort. Most importantly, it has the potential to enable a greater level of input from occupants on the management of their personal environment and can serve as a systematic channel for addressing issues with IEQ related to performance. Finally, leveraging detailed feedback data across multiple projects can help enable evidence-based guidance for the AEC community in the development of more energy efficient, granular and responsive control strategies in line with achieving the dual objectives of low energy performance and enhanced IEQ.

References

American Society of Heating and Refrigeration Engineers (ASHRAE) (2013) ANSI/ASHRAE/ USGBC/IES Standard 90.1-2013. American Society of Heating and Refrigeration Engineers, Atlanta, GA

American Society of Heating and Refrigeration Engineers (ASHRAE) (2014) ANSI/ASHRAE/ USGBC/IES Standard 189.1-2014. American Society of Heating and Refrigeration Engineers, Atlanta, GA

Apte J et al (2006) Window Related Energy Consumption in the US Residential and Commercial Building Stock, LBNL-60146

Brainard GC, Hanifin JP, Greeson JM, Byrne B, Glickman G, Gerner E, Rollag MD (2001) Action spectrum for melatonin regulation in humans: Evidence for a novel circadian photoreceptor. J Neurosci 21:6405–6412

California Energy Commission (CEC) (2013) Title 24 2013 Building Energy Efficiency Standards for Non-Residential Buildings. California Energy Commission, Sacramento, CA

EIA (2012) Commercial Building Energy Consumption Survey (CBECS). Table 6. Electricity consumption by end use, 2012

Franconi E, Huang YJ (1996) Shell, System, and Plant Contributions to the Space Conditioning Energy Use of Commercial Buildings. In: Proceedings for the ACEEE 1996 Summer Study on Energy Efficiency in Buildings, Asilomar Conference Center, Pacific Grove, CA. Washington, D.C.: American Council for an Energy-Efficient Economy

Klepeis NE, Nelson WC, Ott WR, Robinson J, Tsang AM, Switzer P, Behar JV, Hern S, Engelmann W (2001) The National Human Activity Pattern Survey (NHAPS): A Resource for Assessing Exposure to Environmental Pollutants. J Expos Analysis and Environ Epidem 11(3):231–252

Lee E et al (2009) Innovative Façade Systems for Low Energy Commercial Building Systems, https://buildings.lbl.gov/sites/all/files/hpcb-facades-tech-portfolio.pdf

Lockley SW, Brainard GC, Czeisler CA (2003) High sensitivity of the human circadian melatonin rhythm to resetting by short wavelength light. J Clin Endocrinol Metab Sep, 88(9):4502–4505

Lucon O, Ürge-Vorsatz D, Zain Ahmed A, Akbari H, Bertoldi P, Cabeza LF et al (2014): Buildings. In: Climate Change 2014: Mitigation of Climate Change. Contribution of Working Group III to the Fifth Assessment Report of the Intergovernmental Panel on Climate Change [Edenhofer O, Pichs-Madruga R, Sokona Y, Farahani E, Kadner S, Seyboth K et al (eds)]. Cambridge University Press, Cambridge, United Kingdom and New York, NY, USA

New York Times (2012) Crank it Up. Published August 18, 2012.http://www.nytimes.com/ interactive/2012/08/19/sunday-review/19rosenthal-ch-int.html?action=click&contentCollection= Sunday%20Review&module=RelatedCoverage®ion=EndOfArticle&pgtype=article&_r=0. Last accessed 8/29/2016

Shehabi A, DeForest N, McNeil A, Masanet E, Lee ES, Milliron D (2013) US energy savings potential from dynamic daylighting control glazings. Energy Build 66:415–423

Thapan K, Arendt J, Skene DJ (2001) An action spectrum for melatonin suppression: Evidence for a novel non-rod, non-cone photoreceptor system in humans. J Physiol 535:261–267

U.S. DOE (2011) U.S. DOE Buildings Energy Data Book, Table 1.1 Buildings Sector Energy Consumption

U.S. (2016) Green Building Council. LEED v4 for Building Design and Construction. April 5, 2016. 161 pages.http://www.usgbc.org/sites/default/files/LEED%20v4%20BDC_04.05.16_current.pdf

Zelinski EL, Deibel SH, McDonald RJ (2014) The trouble with circadian clock dysfunction: multiple deleterious effects on the brain and body. Neurosci Biobehav Rev, Volume 40, March 2014, pp. 80–101http://www.ncbi.nlm.nih.gov/pubmed/24468109

Chapter 2
The Role of Metrics in Performance-Based Design

2.1 Introduction

To evaluate the performance of buildings in use and to predict performance during design, it is necessary to identify what the appropriate measures of performance should be, when and how measures should be collected, and how results will be interpreted to determine success or failure. As noted in Chap. 1, one of the central barriers to effective daylighting is that daylighting *performance* is often defined differently by different stakeholders, leading to a fragmented approach to performance assessment in the design and operational life-cycle of buildings. While daylighting has most consistently been promoted as a means of electrical lighting energy reduction, greater understanding of the health benefits of daylight and views combined with greater awareness of discomfort glare and the mandate to minimize heating/cooling loads to achieve low and Zero Net Energy (ZNE) buildings has led to an expanding set of performance considerations. Existing metrics, performance criteria and methodologies for assessing daylit spaces have evolved largely from the legacy of metrics developed for the electrical lighting industry, and hold many of the same underlying assumptions. Therefore, it is import to identify and understand the strengths, weaknesses and limitations of these assumptions in assessing the dynamic qualities of daylight spaces.

There are many indicators available for design teams to predict and assess the outcome of daylighting strategies, each with underlying assumptions for the lighting needs, preferences and behaviors of building occupants. This chapter presents a broad assessment of the energy and human-factors performance metrics that should be considered to achieve effectively daylit buildings. The chapter is divided into sections, each discussing a key performance objective. The chapter prioritizes discussion of metrics implemented in consensus-based green building rating systems, whole-building energy benchmarking frameworks and targets, and in software-based evaluation tools on the basis that these metrics are anticipated to have the greatest short-term influence on the design and evaluation of daylit buildings.

© Springer International Publishing Switzerland 2017
K. Konis and S. Selkowitz, *Effective Daylighting with High-Performance Facades*, Green Energy and Technology, DOI 10.1007/978-3-319-39463-3_2

Discussion of each performance objective concludes with an assessment of the limitations with existing approaches and potential opportunities for improvement.

2.2   Optimizing Energy in High-Performance Daylit Buildings

A number of energy efficiency and GHG reduction goals have been developed to transform the design and operation of buildings into an effective tool for mitigating climate change. In the United States, the most ambitious effort is the State of California Long-Term Energy Efficiency Strategic Plan, which has developed and is now striving to implement a vision for all new commercial construction to be Zero Net Energy (ZNE) by 2030 and for 50% of existing commercial buildings to be retrofit to achieve deep levels of energy reduction to achieve ZNE with the addition of clean distributed power generation (CPUC 2008). California also has even more challenging carbon goals to achieve by 2050. A ZNE building is generally defined as, "an energy-efficient building where, on a source energy basis, the actual annual delivered energy is less than or equal to the on-site renewable exported energy" (U.S. DOE 2015). A critical assessment of various ZNE definitions can be found in (Torcellini et al. 2006). While the intent of ZNE is good, we note that in dense urban areas with high rise commercial buildings incorporating energy intensive functions, e.g. data centers, hospitals, etc. it may be physically impossible to meet the requirement with on-site renewables alone.

The emergence of low and ZNE performance goals (discussed in Chap. 1) has placed effective daylighting at the core of whole-building energy efficient design. In commercial buildings, which account for roughly half of the energy used by all U.S. buildings, decisions related to fenestration affect the majority of energy end uses and are thus a central area of focus for performance improvements aimed at enabling low and ZNE buildings. While there are numerous variations on the definition of ZNE, Fig. 2.1 illustrates the general concept. Zero Net Energy is often assessed using the metric of Energy Use Intensity (EUI), which is typically calculated by dividing the total site (or source) energy consumed by the building in one year (measured in kBtu, GJ, or kWh) by the total gross floor area of the building (measured in ft^2 or m^2).

The transmission of daylight through windows (i.e. sidelighting) as a strategy for energy reduction is based on a simple concept: daylight is a renewable light source of high luminous efficacy, which makes the daylighting of buildings an attractive energy strategy compared to the standard practice of constant electrical lighting. As noted in Chap. 1, in the United States, lighting represents the single largest commercial building electricity end use (0.78 exajoules (EJ)) (724 Trillion Btu) (EIA 2012), and is consumed primarily during daylight hours. Of the total averages, it is estimated that 60% is consumed in perimeter zones[1] located 0–12.2 m (0–40 ft) from the building facade during typical daytime work hours (8:00–18:00) (Shehabi et al. 2013).

---

[1]Excluding non-applicable floor space such as religious worship or vacant space.

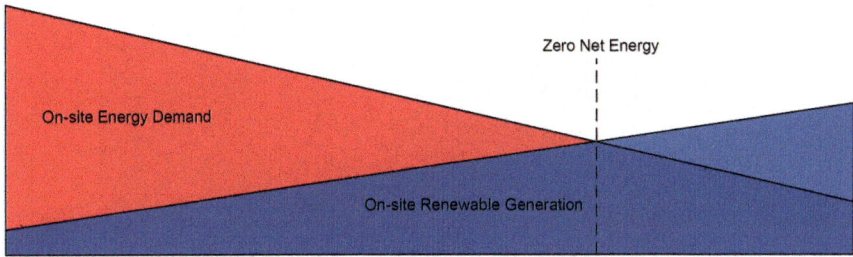

Fig. 2.1 A Zero-Net Energy (ZNE) building generates at least as much energy as it consumes annually

Cooling loads represent another significant energy end use (14%), and one-third is due to electrical lighting and another one-third to solar heat gains through windows (Huang and Franconi 1999). And, because ZNE projects often implement passive or low-energy cooling alternatives such as radiant systems or exposed thermal mass with night-flush ventilation, effective solar control is an additional requirement to avoid exceeding the cooling capacities of these systems, which are typically lower than mechanical HVAC, and consequently more sensitive to peak solar heat gains. Consequently, fenestration strategies that control solar loads and manage glare while transmitting sufficient daylight to minimize the need for electrical lighting in perimeter zones have the potential to significantly improve energy performance.

Figure 2.2 compares the Energy Use Intensity (EUI) of an average office building in Seattle to the recently constructed Bullitt Center (Fig. 2.3), a 6-floor, 4645 m^2 office building designed to achieve ZNE on an annual basis using electricity generated from a rooftop solar photovoltaic (PV) array (a case study

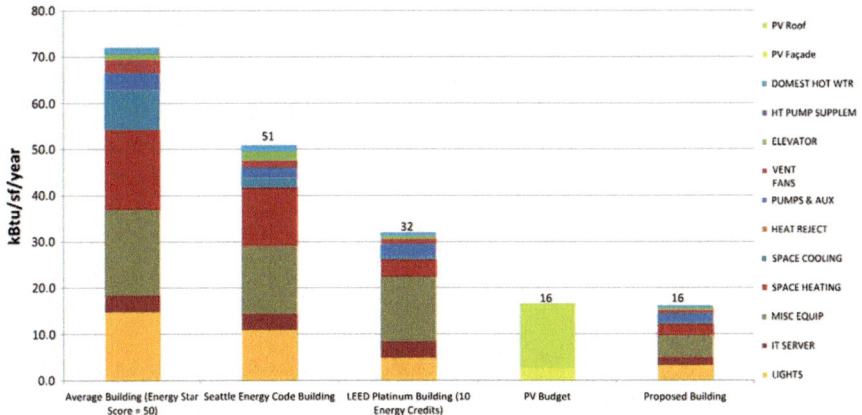

Fig. 2.2 Comparison of the Seattle Bullitt Center (at 50.5 kWh/m^2/yr (16 kBtu/ft^2/year) with various EUI benchmarks and with the renewable energy available from its own rooftop photovoltaic array. *Image credit* The Miller Hull Partnership

Fig. 2.3 Exterior view of the Bullitt Center showing rooftop solar PV array and exterior automated shading devices

description of the Bullitt Center is provided in Chap. 5). Due to the spatial constraints of the project site, the available area for a PV array on the roof combined with the relatively cloudy Seattle climate led to a renewable energy "budget" of 50.5 kWh/m^2/yr (16 kBtu/ft^2-yr). Compared with an average Seattle office building, or even a LEED Platinum office building, the PV budget was found to meet only 22 and 50% of those annual energy requirements respectively (Fig. 2.2). Driven by the spatial constraints of the site, local climatic conditions, and the ZNE performance target, the design team worked to develop a highly efficient building envelope to minimize loads and enable the application of passive environmental control strategies of daylighting, direct gain solar heating, natural ventilation, and night-flush cooling. These strategies were combined with low-energy mechanical systems (ground source heat pumps, in-floor radiant heating/cooling, and a Dedicated Outdoor Air System (DOAS) with heat recovery, and resulted in a designed EUI that could be met by the renewable energy budget of 50.5 kWh/m^2/yr (16 kBtu/ft^2-yr).

Daylighting, a thermally efficient envelope, and actively managed fenestration systems are key components of the Bullitt Center whole-building energy efficiency strategy. Figure 2.2 shows the result of daylighting on electrical lighting EUI, where daylight effectively reduces the operational hours of electrical lighting.

Similarly, automated facade shading acts as a dynamic filter to enable both passive solar heating and solar shading when required to significantly reduce space heating and cooling loads.

2.2.1 From Daylight "Harvesting" to Daylight Autonomous Buildings

While EUI is perhaps the singular most consensus-based metric for gauging energy efficient design, human factors ultimately determine the long-term viability of design strategies, and serve as the underlying basis for differentiating energy "use" from simply energy consumption or waste. Figure 2.4, which presents an interior view of the perimeter zone workspaces within the Bullitt Center, illustrates how a project designed to achieve a low EUI was able to maintain relatively large areas of facade glazing in heating dominated climate, preserving daylight access and views for occupants. In Fig. 2.4, it is important to note the absence of a conventional installation of direct/indirect electrical lighting fixtures on the ceiling. As part of the overall ZNE goal, the client decided to install minimal fixtures and require tenants to install supplemental electrical lighting if desired as a tenant improvement. On site observations revealed that no tenants have installed additional electrical lighting as of the publishing of this book. Because tenants were satisfied with the lighting

Fig. 2.4 Sixth floor daylit perimeter zone. *Note* the absence of a conventional installation of electrical lighting fixtures on the ceiling

conditions provided by daylight, (supplemented with task lighting when necessary), the installed electrical lighting power demand is extremely low, or zero in many spaces.

This outcome is notable, as it indicates that the perimeter zones effective operate as "daylight autonomous" spaces from the perspective of their occupants. In conventional energy efficient lighting design practices, daylit perimeter zones would be designed with photocontrolled (i.e. daylight-dimming) ambient lighting systems to "harvest" available daylight, rather than be considered as zones that should require no installed ambient lighting. The Bullitt Center provides a glimpse of the potential for perimeter zones to be classified as "daylight autonomous," where significant energy and cost savings can be achieved through the minimization or elimination of supplemental electrical lighting within 4 or 5 m (13–16.5 ft) from the facade. The critical lesson is that the performance of the architectural daylighting strategy and resulting occupant-based daylight availability should be evaluated prior to the consideration of technology solutions that may reduce energy, but be viewed as unnecessary from the perspective of building occupants.

At any point in time, codes and standards for lighting dictate much of what is designed and built. Practitioners often assume that the current standard practice is somehow optimized, up to date and reflecting immutable norms that persist over time. In fact, "best practice" in design is in flux all the time, although normally changing slowly since change is always a challenge. A lighting design, completed this year, captures and embodies (1) the owners preferences for what they want for their staff or they think the market wants or will accept, (2) mandatory and voluntary codes and ratings constraints, (3) what the design team can reasonably deliver on time and budget with minimal risk, (4) what competing manufacturers can deliver in volume to a job site, (5) what contractors can properly install and commission, (6) operating costs for electricity and (7) what occupants can effectively operate to meet their needs for comfort, health and productivity.

It is not surprising that while there is much diversity in the practice of lighting design, mainstream practice changes slowly. But change does happen and is driven by (1) emerging technology with enhanced, affordable features, e.g. LEDs, sensors, wireless controls, (2) changing demands by owners, e.g. LEED ratings, (3) changing regulatory requirements, e.g. utility demand response programs, stricter state building codes, (4) economic pressures of operating costs, and (5) new knowledge and perspectives about occupant performance, needs and preferences with respect to lighting.

Less than 100 years ago daylight was the preferred and primary source of lighting in many buildings. The advent of the electric light, and particularly the fluorescent lamp, and the growth of an electric infrastructure to deliver power to every building transformed the design of lighting in buildings. Permanent Supplementary Artificial Lighting (PSALI) was the new invention at the time, based on the underlying novel concept that one could rely on electric lighting as the primary light source rather than daylight. This rapidly became the norm for office design and as the cost of delivered electric lumens fell, office lighting standards suggested uniformly high illuminance levels throughout occupied spaces, and even operating the lights 24 hours a day to provide heating. In an era when low and high pressure lamps were improving in output and color, and when nuclear power promised electricity that would be "too cheap to meter" this vision of building lighting became the norm.

While lamp technology continued to improve, the availability of cheap, reliable energy ended abruptly in the 1970s, first driven by availability, and later by environmental concerns related to carbon emissions. In that context 'daylight" was rediscovered as a strategy to reduce reliance on electric lighting by simply reducing output to lights when the resource was available. The design skills of 50 years earlier in terms of how to size and manage fenestration to admit daylight without glare or solar load were rediscovered, reinvented and improved upon, as was the lamp, sensors and controls infrastructure needed to capture the potential savings. But these changes never made it into the mainstream of practice and are just now being mandated by some building codes, the last step in the process of more widespread adoption. We are now 40+ years into that new cycle of change and once again technology, i.e. new efficient light sources and the Internet of Things, is driving some of that change. But major new forces on "best practice" is also being driven by a renewed interest in the role of lighting and daylighting on occupant health, well being, comfort and performance, factors that were often overlooked, forgotten or ignored in the past. In this new context there are exciting changes in play in the design landscape for lighting and daylighting design.

As design goals shift from electrical lighting energy "savings" towards efforts to optimize the potential of daylighting within a whole-building energy concept, reliable performance indicators and methods for assessing daylight sufficiency during design are needed. While assessments of EUI, peak demand, and peak cooling loads are critical for meeting carbon reduction targets, demand side load management, and for enabling the application of low-energy cooling technologies in daylit buildings, emerging metrics for assessing daylight sufficiency are critical for optimizing energy goals around end-user needs and preferences for daylit environments. The following section frames emerging research in Climate Based Daylight Modeling (CBDM) and associated metrics as an effort to improve the ability of designers to deliver daylight autonomous buildings.

2.3 From Static to Dynamic, Climate-Based Daylighting Metrics

As designers seek to go from simply "maximizing" daylight through architectural transparency to thoughtfully managing the admission of daylight to address explicit programmatic and occupant needs within the limits of local climate and building energy goals, new metrics that are sensitive to the unique, time-varying daylighting conditions of the project site and local climate are needed. Historically, the daylight factor (DF) was the most widely applied metric used to assess daylight sufficiency (Nabil and Mardaljevic 2005). The daylight factor is defined as the ratio of the internal illuminance at a point in a building to the unshaded, external horizontal illuminance under a CIE overcast sky, (Moon and Spencer 1942). It originated in Europe as a metric to assess the daylight conditions needed to provide minimally "adequate" daylight levels. Since the worst case conditions i.e. minimum daylight levels, were overcast skies, that was used as the basis for analysis. Use of the DF method is common because it is simple to understand and relatively easy to measure, leading to its use in codes and standards in the UK and Europe. Over time DF began to be used as a metric to assess annual performance, which it is poorly equipped to do because it does not account for clear skies, partly cloudy conditions or direct sunlight.

The use of DF as a metric to assess daylighting performance has been further compromised because the absolute values selected as design targets have not always been well thought out. In previous versions of LEED, (e.g. USGBC 2009), an average DF of 2% across a given space was required for it to be considered sufficiently daylit. Since it did not account for direct sun conditions the actual daylight values in spaces could be much higher. In addition because it is based on assessments of horizontal illuminance under standard overcast sky conditions, it is not sensitive to building orientation, geographic location, sun position, or daily/seasonal changes in sky conditions. This is particularly problematic for projects that are located in climates where standard overcast skies rarely exists, and for low and ZNE projects where assessing solar control is a critical design factor. Second, because the DF does not account for the effects of direct beam radiation, and because there is no consensus for an acceptable "upper limit" for the ratio, the DF approach has been criticized for incentivizing a "the more transmission the better" approach, where spaces that would have uncomfortable direct sun or glare can not be differentiated. Finally, the DF approach does not easily allow for the evaluation of aspects of the design that may respond dynamically to changes in weather or sun position, such as automated facade systems or interior shading devices, which are increasingly common in low and ZNE projects.

To address the overly simplified static approach of the DF more complex hourly daylight simulation models were developed beginning in the 1980s. These started with the geometric design of the space to be modeled, utilized the optical properties of glazing and shading systems and calculated the interior daylight levels at several locations in the room using a variety of methods for given latitude, time of year, hour of the day, and weather conditions (Ward 1998). To determine annual energy impacts simplified versions of these models were embedded in hourly simulation programs such as DOE-2, which then calculated daylight levels at several control points in a space on an hour by hour basis using location-specific hourly weather files (Selkowitz et al. 1982). These tools provided hourly illuminance data at control points throughout the year that were climate dependent, location dependent and orientation dependent, and could accommodate the deployment of shading systems, and also calculated simplified glare indices on an hourly basis. These hourly data were primarily used to estimate annual lighting energy savings and overall building energy performance with a focus on the daylighting solution as an energy saving strategy. Most of the 8760 hourly calculations were completed using coefficient modeling approaches due to the computational intensity of the more accurate first principles calculations and the limitations of the widely used desktop computing systems.

Increased interest in the "subtleties" of daylighting performance coupled with improved models and more computational power now provide more options to determine climate specific data on a more granular spatial and temporal scale. To enable the dynamic, time-varying attributes of a project and its climate to be more fully evaluated, researchers have further developed an approach now generally referred to as Climate Based Daylight Modeling (CBDM) (Mardaljevic 2006). Climate Based Daylight Modeling involves the prediction of interior daylighting conditions over an annual period using sky models derived from standardized hourly weather data representative of the project location. The benefit of CBDM is that it enables designers to develop projects in response to the unique solar and weather conditions of the project site as well as to more readily implement dynamic changes in daylight apertures such as automated facade systems deployed for direct sun control or manually operable interior shading devices deployed to reduce glare. Figure 2.5 presents a comparison of annual hourly weather information (global horizontal illuminance (lux)) between Stockholm, Sweden and Phoenix, Arizona that illustrates differences in the seasonal availability and intensity of daylight. By examining daily and seasonal changes in the spatial patterns and intensity of daylight, designers can predict where and when designs perform well or poorly in regard to daylight sufficiency as well as the potential for glare and solar over-heating. The outcome of a thoughtful design process utilizing CBDM is a unique design solution tuned to the local site. While CBDM provides more site-specific quantitative prediction of daylit illuminances achieved by a particular design option, the approach introduces a number of additional considerations, such as what amounts of daylight are considered insufficient or sufficient by occupants, how dynamic changes in light should be assessed spatially and on an annual basis, and what conditions are likely to be associated with glare and the operation of shading devices. The following section discusses the procedures and metrics used in CBDM and concludes with a discussion of limitations and needs for further research.

Because the lighting conditions in a DF assessment are static, the calculation procedures require only a single, "point-in-time" assessment of points with a space and can be achieved in relatively short time using many computational methods including raytracing programs (e.g. Radiance). In contrast, CBDM is a temporal assessment of lighting conditions over a specified time interval. For example, an annual assessment on an hourly basis (assuming daylight hours from 6:00–18:00) would require 4380 unique assessments. Consequently, CBDM developed along with improvements in computing capacity and changes to software simulation approaches. Due to the significant computational time required by raytracing methods, CBDM utilizes the daylight coefficient approach originally developed by

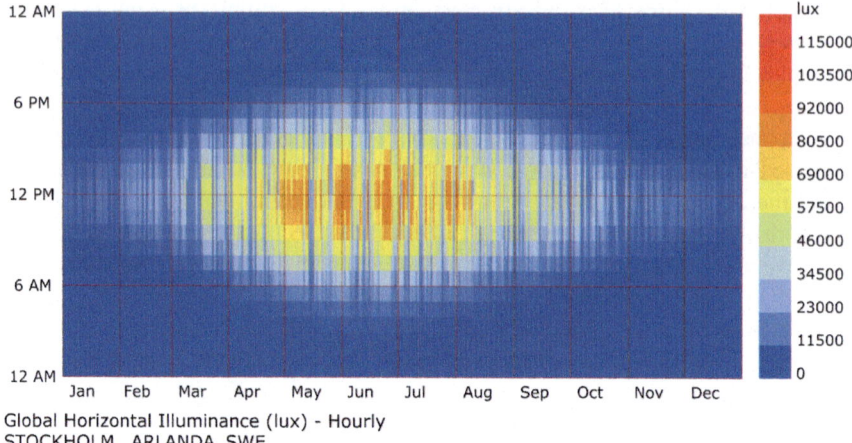

Global Horizontal Illuminance (lux) - Hourly
STOCKHOLM_ ARLANDA_SWE

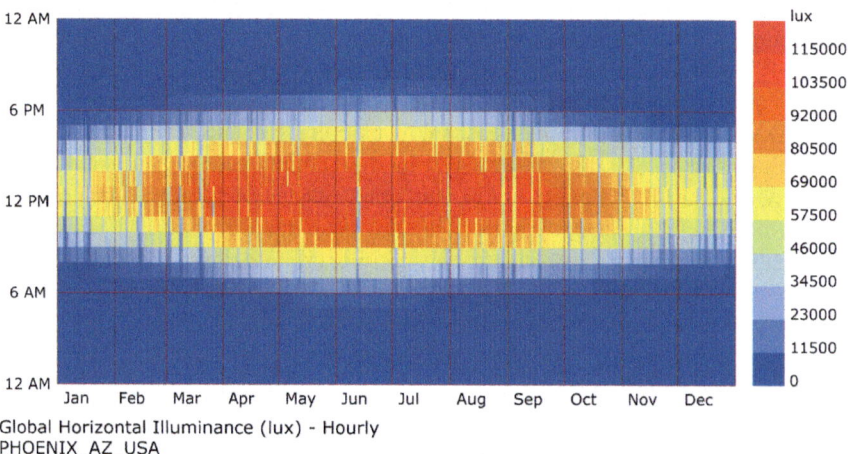

Global Horizontal Illuminance (lux) - Hourly
PHOENIX_AZ_USA

Fig. 2.5 Comparison of availability and intensity of hourly global horizontal illuminances (lux) for Stockholm and Phoenix illustrating differences in the seasonal availability and intensity of daylight between climates

Tregenza and Waters (1983) and first implemented using *Radiance* tools by Reinhart and Herkel (2000) and later standardized by (Bourgeois et al. 2008).

In contrast to conventional raytracing methods, the daylight coefficients for a given point do not depend on the luminance distribution of the sky vault. They are only dependent on the building geometry, aperture dimensions and optical characteristics, interior surface characteristics, and the sub-division of the sky and ground into a matrix of patches (Figs. 2.6 and 2.7). Once the coefficient for each patch has been calculated, an algebraic equation can be used to determine the illuminance at a point, given an arbitrary sky distribution. This approach significantly reduces

Fig. 2.6 Example Tragenza
sky matrix consisting of 145
"patches" displayed over a
hypothetical sidelit space

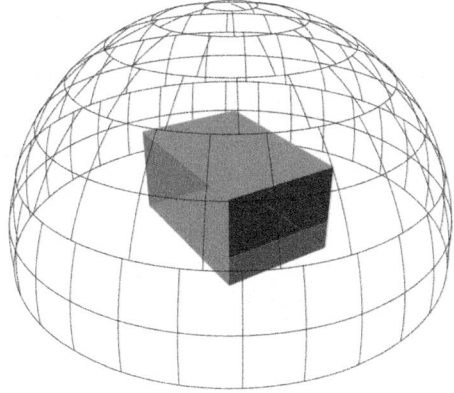

Fig. 2.7 *Top view* of
Tragenza sky matrix

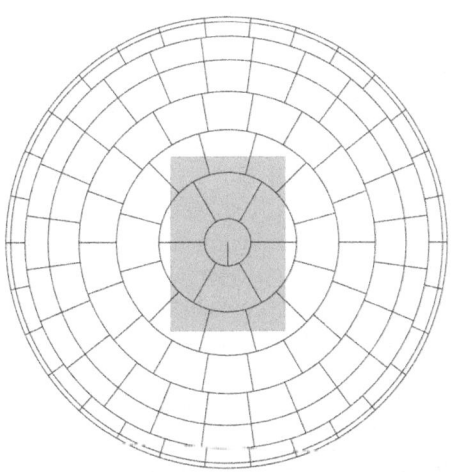

simulation time, enabling the analysis of annual daylighting conditions over short
time steps (e.g. hourly). Consequently, the contribution of daily and seasonal
changes in daylight availability, direct sun penetration, and glare can be assessed,
either cumulatively for the full year, or on an hourly basis while maintaining rea-
sonable photometric accuracy for simple fenestration systems.

2.3.1   Climate-Based Daylighting Performance Metrics

Daylight levels in spaces range over several orders of magnitude i.e. from 10 to
100,000 lx, within a single space over time and weather conditions so metrics that
distinguish time dependent effects, upper and lower limits and spatial effects are

all potentially important for design. Several metrics have been proposed to evaluate performance using the CBDM approach. These include *Daylight Autonomy* (DA) (Reinhart 2002), *Useful Daylight Illuminance* (UDI) (Nabil and Mardaljevic 2005), *Continuous Daylight Autonomy* (CDA) (Rogers 2006), and *Spatial Daylight Autonomy* (sDA) (IES 2012). Daylight Autonomy was originally defined by Reinhart as: "The percentage of occupied times of the year when a minimum work plane illuminance threshold of 500 lx can be maintained by daylight alone." The DA metric is used to indicate the percentage of occupied hours of the year when daylight is sufficient to eliminate the need for electrical lighting. Based on a concern that the binary threshold approach of the original DA criteria artificially differentiated between spaces that may not be perceived as different by the human visual system (e.g. 470 vs. 510 lx), Rogers (2006) proposed the CDA metric, assigns a fractional weighting to illuminances below the established threshold in the annual summary of daylight availability. The original DA criteria were expanded by Nabil and Mardaljevic (2005) in their UDI metric to include a "discomfort" threshold of 2000 lx, and reduced the minimum daylight illuminance threshold to 100 lx. The authors note that these limits are based on reports of occupant preferences and behavior in daylit offices with user-operated shading devices. Occupied hours of the year where the horizontal illuminance does not fall within these limits (100–2000 lx) are omitted from the annual summation of UDI.

The IES Approved Method for sDA and ASE (LM-83) is an attempt to define a standardized calculation and simulation-based modeling methodology to predict daylighting performance. Spatial Daylight Autonomy (sDA) is defined as the percent of an analysis area that meets a minimum horizontal daylight illuminance level (e.g. 300 lx) for a specified fraction (e.g. 50%) of the operating hours per year (IES 2012). It is written using the subscript $sDA_{300,50\%}$. The basis for the illuminance thresholds and performance criteria is largely derived from field research, which consisted of measured data and expert assessments conducted in 61 buildings (Heschong 2012). An sDA outcome is calculated as the percent of analysis points across the analysis area that meet or exceed the 300 lx threshold for at least 50% of the analysis period and is reported a single number ranging from 0 to 100%. The analysis period is from 8:00AM to 6:00PM each day, including weekends, leading to 3650 h per year, regardless of building type, space use (i.e. program), or project location on the earth (e.g. latitude). The IES has defined two performance criteria based on sDA outcomes, "Preferred" and "Nominally Accepted." Analysis areas must meet or exceed $sDA_{300,50\%}$ over 75% of the analysis points to be rated as "Preferred." Analysis areas must meet or exceed $sDA_{300,50\%}$ over 55% of the analysis points to be rated as "Nominally Accepted."

Unlike UDI, sDA has no upper limit on daylight illuminance. Therefore, to evaluate the potential risk of excessive sunlight penetration, the IES daylighting metrics committee developed an accompanying metric entitled Annual Sunlight Exposure (ASE). ASE is a metric that, "describes the potential for visual discomfort in interior work environments" (IES 2012). ASE is calculated using the same analysis points and analysis period as sDA and quantifies the percentage of analysis points that receive at least 1000 lx for at least 250 occupied hours per year. It is written using the subscript $ASE_{1000,\ 250h}$. There are three performance criteria for evaluating excessive sunlight penetration based on $ASE_{1000,\ 250h}$ outcomes. Daylit spaces predicted to have more than 10% $ASE_{1000,\ 250h}$ are considered to have "unsatisfactory visual comfort," spaces with less than 7% are considered "nominally acceptable" and spaces with less than 3% are considered "clearly acceptable" (IES 2012). Notably, the $sDA_{300,50\ \%/}$ ASE simulation method is the first to attempt to standardize the inclusion and operation of interior shading devices to control direct sun in order to present a more realistic prediction of daylight availability in zones considered to have glare and significant periods of direct sun penetration. The simulation method requires that all exterior windows must be modeled with interior shading devices unless the zone associated with the window is determined to be "nominally" or "clearly acceptable" based on $ASE_{1000,\ 250h}$.

Figure 2.8 presents an example evaluation using $sDA_{300,50\ \%}$ and $ASE_{1000,\ 250h}$ of a daylit space with a high level of facade glazing on two elevations. The space is 12 m by 12 m by 3 m in size and the glazed facades are oriented N and E respectively. The project is located in Pasadena, CA (34.15 N latitude), a climate dominated by clear skies and direct sun. Figure 2.8 presents a perspective (upper image) and plan view (lower image) of the $ASE_{1000,\ 250h}$ analysis. The analysis grid has been enlarged slighting from the recommended maximum spacing (2ft., 0.61 m) to a grid spacing of 0.75 m for illustrative purposes. The $ASE_{1000,\ 250h}$ analysis shows that the space receives direct sun over a large fraction of the analysis grid and the result of 50% is significantly above the threshold indicating that "unsatisfactory visual comfort" is likely (>10%).

Following the LM-83-12 modeling methodology,[2] interior window shades and shade operating behavior are included in the calculation of $sDA_{300,50\%}$. Figure 2.9 shows the resulting shading profile for the east-facing window group and Fig. 2.10 shows the shading profile for the north-facing window group. Due to the predominately sunny Pasadena climate, the inclusion of shades which deploy on an hourly basis when more than 2% of analysis points within the window group exceed 1000 lx leads to an east facade that is completely shaded during daylight hours and a north facade that is shaded for a significant number of morning hours.

Figure 2.11 shows the resulting $sDA_{300,50\%}$ outcome for the space. The same grid spacing is used as in Fig. 2.8. As a result of the extensive window shading on the east facade, the region of the space that achieves the greatest levels of DA are oriented towards the north windows, and the contribution of the east facade to interior daylight is minimal, despite the significantly greater amount of solar radiation incident on the facade exterior.

---

[2]See IES (2012) for a complete description of the climate modeling methodology.

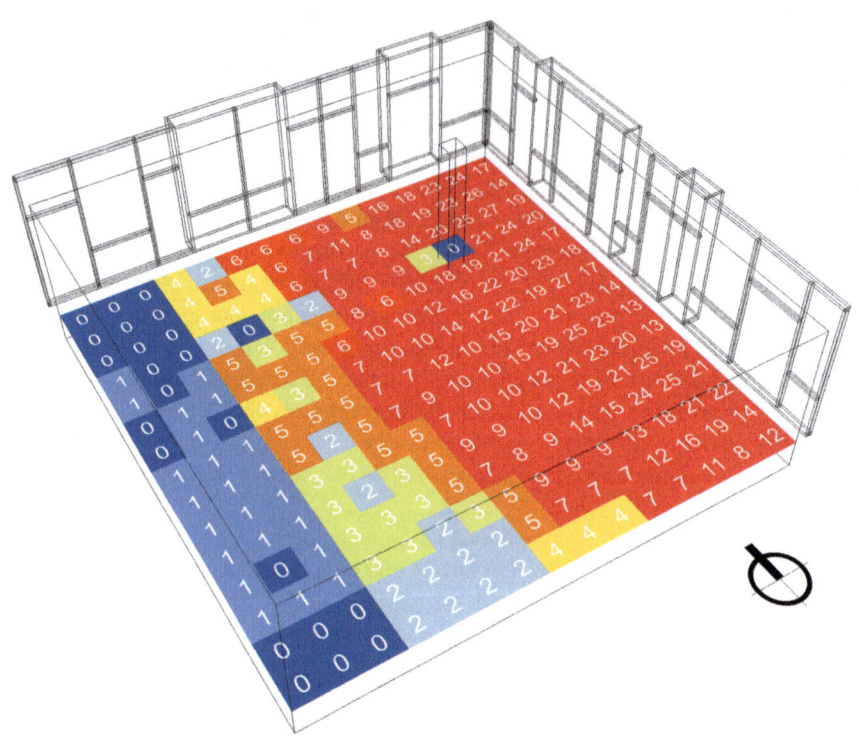

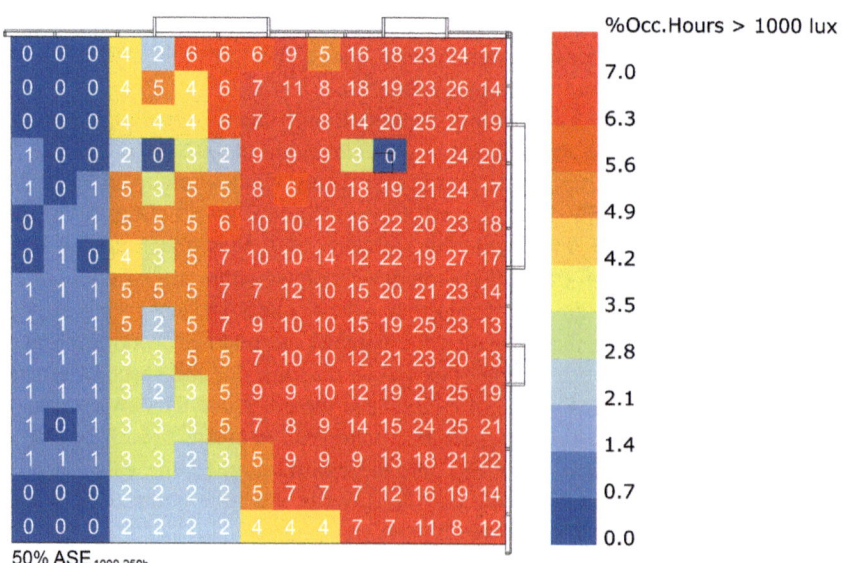

50% ASE$_{1000,250h}$

%Occ.Hours > 1000 lux

| 7.0 |
| 6.3 |
| 5.6 |
| 4.9 |
| 4.2 |
| 3.5 |
| 2.8 |
| 2.1 |
| 1.4 |
| 0.7 |
| 0.0 |

◀**Fig. 2.8** Perspective view (*upper image*) and plan view (*lower image*) showing $ASE_{1000,\ 250h}$ result. Each *square* indicates the percentage of occupied hours of the year where the *square* exceeds the illuminance threshold of 1000 lx, a threshold indicator for the presence of direct sun that may cause discomfort. Based on an analysis period of 3650 h per year (10 h per day (8:00AM to 6:00PM) and includes weekends). *Squares* in *red* indicate regions that exceed 1000 lx for 7% (250 h) or more of the 3650-hour analysis period

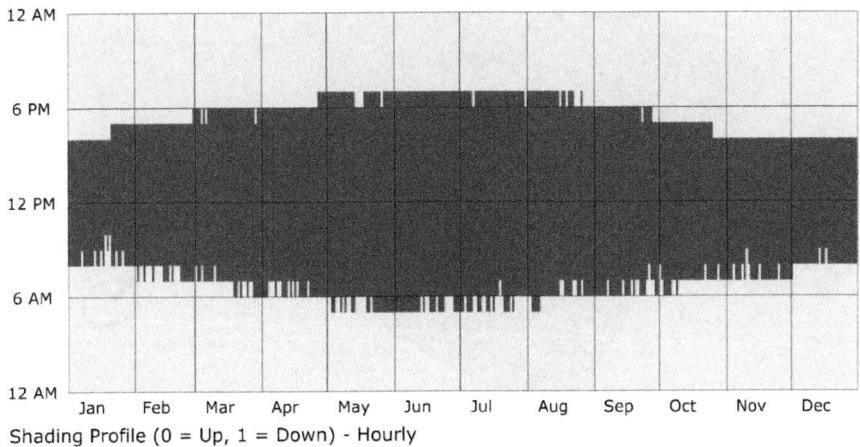

Shading Profile (0 = Up, 1 = Down) - Hourly

Fig. 2.9 East facade hourly shading profile (*dark grey* color indicates shade deployed)

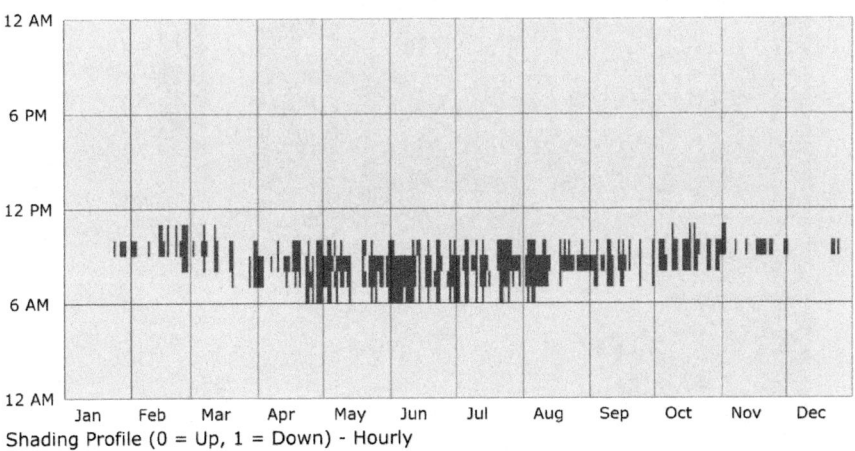

Shading Profile (0 = Up, 1 = Down) - Hourly

Fig. 2.10 North facade hourly shading profile (*dark grey* color indicates shade deployed)

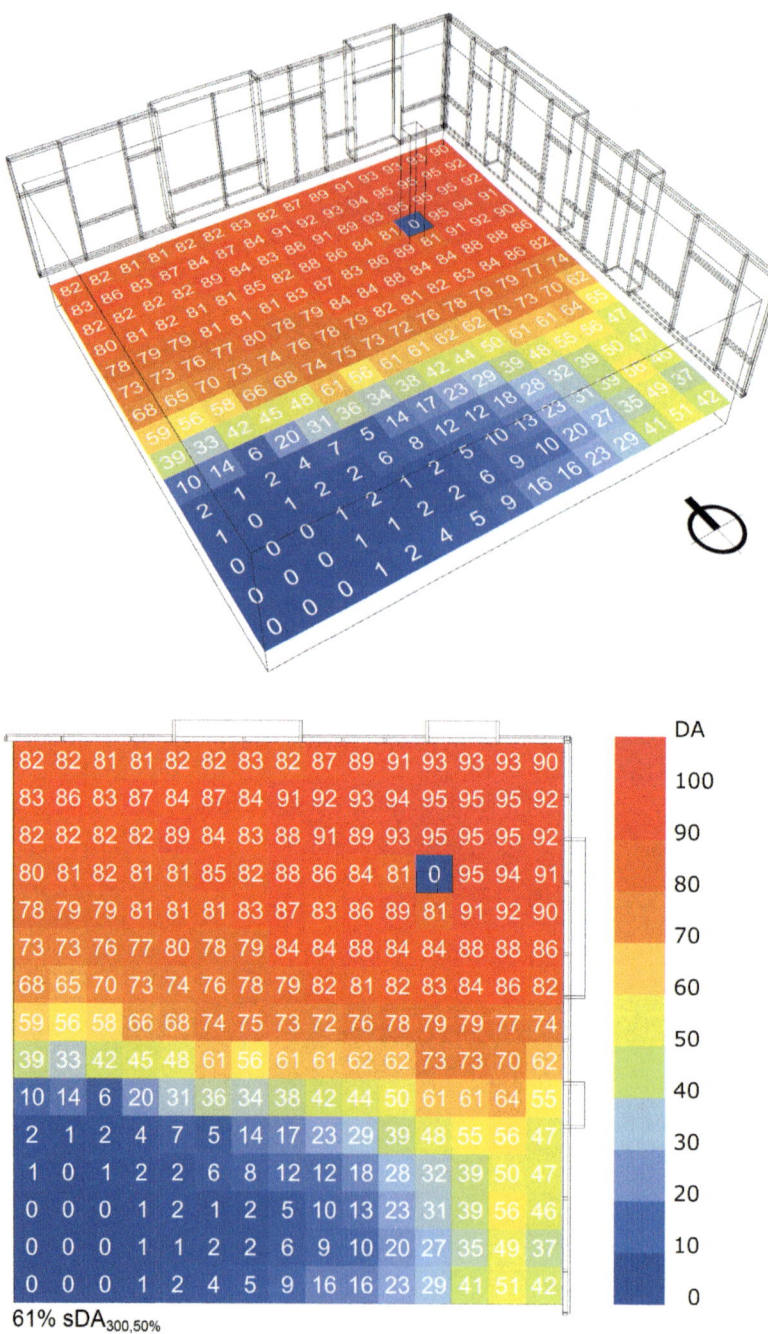

Fig. 2.11 Perspective view (*upper image*) and plan view (*lower image*) showing $sDA_{300,50\%}$ result. Each square indicates a unique Daylight Autonomy value, the percentage of occupied hours of the year (0-100%) where the *square* exceeds the illuminance threshold of 300 lx, a threshold indicator for sufficient daylight. The number of *squares* that equal or exceed 50% (138) are divided by the zone total (225) to determine $sDA_{300,50\%}$ (138/225 = 61%)

2.3.2   *Limitations and Future Directions of Climate-Based Daylight Modeling*

Climate Based Daylight Modeling represents significant progress in efforts to improve the fidelity of simulation-based predictions of daylighting performance. In particular, the IES Approved Method for sDA and ASE (LM-83), makes an ambitious effort to link annual daylighting performance of spaces with the preferences and actions of building occupants through assumptions for shading device use. LM-83 has recently been adopted for use in determining compliance with the LEED Daylighting Credit (U.S.G.B.C. 2015). In addition, LM-83 is referenced in ASHRAE 100-2015 (Energy Efficiency in Existing Buildings), and the analysis method used for mandatory and prescriptive photocontrol requirements for California's energy efficiency standard (Title-24) is based on LM-83. Due to the growing use of these systems and standards, one might assume that the IES Approved Method for sDA and ASE will become the consensus metrics used for predictions and claims of effective daylighting. However, it is important to note that very few projects have applied these new metrics and, largely due to the complexity embedded within the annualized simulation methodology, there is no procedure for directly comparing predictions with performance of built projects in use to validate the embedded assumptions regarding occupant needs and behavior. This is perhaps the most significant general limitation of any annualized simulation-based approach to daylighting evaluation.

It is also important to note that these outcomes do not relate directly to energy outcomes for photocontrolled electrical lighting systems. Although Daylight Autonomy is often applied to indicate regions of the work plane where electrical lighting is not required over a period of time, the assumption that either manual switching or dimming, or that a photocontrolled electrical lighting system will modulate light output in direct proportion to incident daylight at each region represents a theoretical upper limit for energy reduction potential. In practice, discrete lighting zones are generally controlled in a closed loop by one interior photosensor placed at a specific point that is intended to be representative of the zone illuminance and having a single view of some region within the zone. Consequently, to simulate lighting energy reduction, at least one sensor point and view vector must be defined for each zone. The modeling of occupant shade control behavior or automated controls adds an additional layer of complexity, due to a similar need to define the critical view points and view vectors for registering the stimulus condition assumed to drive occupant behavior or to control an automated system.

While CBDM methods represent an improvement from assessment methods of the past they are still an evolving work in progress. The following represent factors that should be considered when applying CBDM and some of the new metrics based on IES LM-83 to a design project:

First, the criteria used to differentiate daylight illuminances acceptable to occupants (e.g. 300 lx, global horizontal illuminance) from levels perceived to be insufficient are not supported by a large body of subjective responses to transient daylighting conditions from buildings in use. Rather, the threshold appears to be used largely due to its legacy as a common standard horizontal illuminance level for electrical lighting design in offices. As visual tasks and office occupancy continue to change these thresholds might change as well.

Second, as suggested by Reinhart (2015), the annual solar exposure limit implemented in LM-83 largely precludes direct sunlight from entering a space, which may be overly restrictive for many space uses where occupants accept (and even prefer) the presence of direct sun. Critical visual tasks in offices may require good control of direct sunlight but there are many workspaces in most building types for which some sunlight penetration may be welcomed, particularly in colder and cloudy climates.

Third, while the human visual system is frequently oriented vertically, sDA and ASE (in addition to all commonly-used daylighting metrics) are derived from measurements of horizontal illuminance on a theoretical "horizontal workplane." This measurement approach is a legacy of lighting research focused on horizontal visual acuity task performance when workers read documents on a desk and is likely to continue to be poorly applicable to predicting occupant perception and appraisal of the luminous environment with emissive vertical displays in a modern, evolving work space. This is particularly of concern in the assessment of glare. With the eye oriented vertically, direct view of the solar disc or extreme luminance contrasts between windows and indoor surfaces can often become sources of glare and visual discomfort which do not correlate with local (e.g. workstation) measurements of horizontal illuminance.

The fourth limitation is the reliance of daylighting metrics on the photometric quantity of illuminance (lumen per m^2), rather than luminance (candela per m^2). While current daylighting metrics focus exclusively on absolute measurements of illuminance incident on often-imaginary horizontal surfaces, the visual system responds to patterns of luminance in the field of view (the amount of light transmitted, emitted or reflected from real surfaces). Further, the perception of glare in a field of view is known to include an adaptation effect and depends on the luminance of the viewed surface relative to other surfaces in the field of view, not simply the absolute luminance of the surface. Therefore, while measures of horizontal illuminance have a long history in human-factors studies of light, alternative approaches are needed that more closely address the contemporary human experience of light in buildings, both in simulation-based environments and in real buildings. Vertical, luminance-based metrics, such as the assessment if Daylight Glare Probability (DGP) for glare, which leverages the luminance-mapping capabilities of High Dynamic Range (HDR) imaging, present one alternative with significant promise.

Finally, the LM-83 simulation methodology assumes that interior shading devices will be fully deployed by occupants in the presence of direct sun and fully retracted when direct sun is not present. Deviations from this "active operator" assumption in real buildings will result in significantly different quantities of illuminance, which form the basis for the daylight autonomy criteria, as well as differences in glare. To support the effective use of daylighting metrics, it is important to develop a body of human factors data from buildings in use that demonstrates a relationship between the performance indicators and subjective assessments of daylight illuminance. It is additionally important to examine the extent to which realistic occupant operation or automated operation of shading devices may increase or decrease predicted daylight availability in buildings in use.

2.4 Non-visual Effects of Light

Standards and practices for lighting design (both daylighting and electrical) in buildings were developed based primarily on pragmatic needs of performing visual tasks but only on a limited scientific understanding of the important role light plays in maintaining healthy human biological functions. In indoor environments, where it is estimated that U.S. adults spend nearly 87% of their lives (Klepeis et al. 2001), lighting is often provided by electrical sources that are adequate for visual task performance, but lack the appropriate spectrum and intensity required to stimulate the circadian system. As described by the Illuminating Engineering Society of North America (IESNA), the formal definition of light is "radiant energy that is capable of exciting the human retina and creating a visual sensation" (IESNA 2016). The recent discovery of a third class of photoreceptors in the human retina (Provencio et al. 2000; Gooley et al. 2001; Hannibal et al. 2002; Hattar et al. 2002), referred to as Intrinsically Photoreceptive Retinal Ganglion Cells (ipRGCs), is serving to add an additional and complex set of new considerations and performance expectations for lighting designers.

The human circadian system (or circadian clock) is responsible for orchestrating the daily timing of physical, mental and behavioral changes. These include sleep/wake, alertness level, mood, hormone suppression/ secretion, and core body temperature (CIE 2004). In the majority of humans, the period of the SCN is slightly greater than 24 h. In order to maintain entrainment with the local 24-hour light/dark cycle, the circadian system relies on a resetting response driven by light received at the retina. The magnitude of the resetting response is dependent on a number of parameters including the timing, intensity, duration, wavelength, number and pattern of light exposures (Lockley et al. 2003). The lack of a sufficient light stimulus at the appropriate time can disrupt the circadian system, which can in turn lead to a range of negative health outcomes, such as poor sleep, reduced alertness, and increased risk of a range of health problems including diabetes, obesity, cardiovascular disease and cancer (Zelinski et al. 2014). Common causes of circadian disruption include long-distance travel, night-shift work, exposure to bright light in the evening, and long-term occupancy in poorly lit indoor environments.

Relative to the visual system, which is maximally sensitive to ($\sim$555 nm) "green" light, the action spectrum of the circadian system is shifted towards shorter wavelength ($\sim$480 nm) "blue" light (Brainard et al. 2001; Thapan et al. 2001). Thus, the photopic luminous efficacy function (V-lambda) and standard photometric units (lux) are problematic for assessing the biological effects of various light sources. Figure 2.12 shows the spectral response function of the circadian system (C-lambda) and the visual system (V-lambda) along with the Spectral Power Distributions (SPDs) of three CIE daylight illuminants (D55) sunlight, (D65)

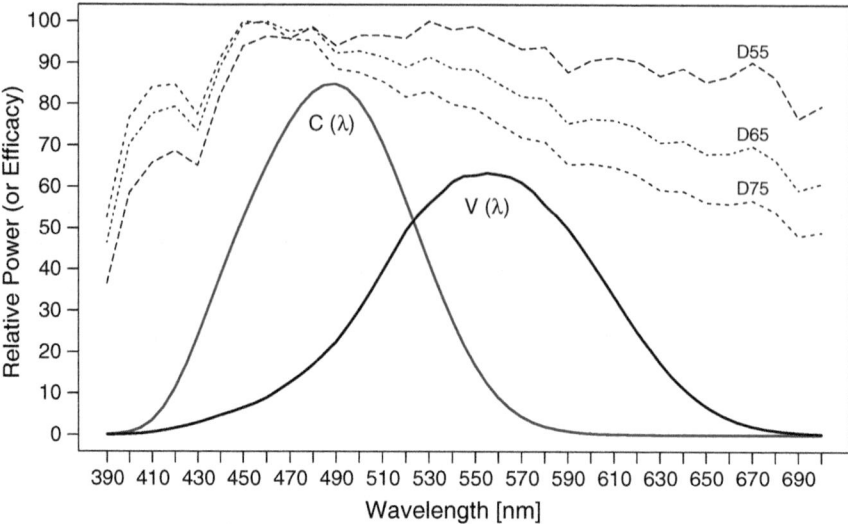

Fig. 2.12 Comparison of spectral response of the visual (photopic) system (V-Lambda) and the circadian system (C-Lambda) to the relative spectral power distributions of three CIE daylight illuminants: (D55) sunlight, (D65) overcast sky, and (D75) north sky daylight. *Note* Both response curves are scaled to have equal area under the curves

overcast sky, and (D75) north sky daylight. It can be seen from Fig. 2.12 that the peak sensitivity of the circadian system (C-lambda) matches closely with the peak power of various daylight SPDs. In contrast, Fig. 2.13 compares the spectral response of the visual system (V-lambda) and the circadian system (C-lambda) to the spectral power distribution a narrow tri-band fluorescent lamp (the CIE illuminant F11) installed in many commercial office building lighting applications. Figure 2.13 shows that the peak power of the fluorescent light aligns closely with the response function of the visual system (V-lambda) and that relatively little power is distributed within the sensitivity of the circadian response function (C-lambda).

Timing of light exposure also plays an important role in synchronizing circadian rhythms with daily patterns of activity (Khalsa et al. 2003). For a typical well-rested and regularly-sleeping individual, a light stimulus in the early morning will advance the circadian clock, causing earlier wake-up time and earlier sleep onset. Alternatively, light received in the evening will delay the circadian clock, causing later wake-up time and later sleep-onset. Light received in the middle of the biological day will have limited effect on circadian advancement or delay, but has been shown to cause reduced levels of sleepiness and higher levels of subjective alertness (Phipps-Nelson et al. 2003; Rüger et al. 2006). Finally, past history of light exposure has an effect on sensitivity of the circadian system to light (Chang et al. 2011). Higher levels of light exposure during the day cause the sensitivity of the circadian system to decrease over time, and lower exposure levels causes sensitivity to increase. A thorough summary of the parameters that control the response of the

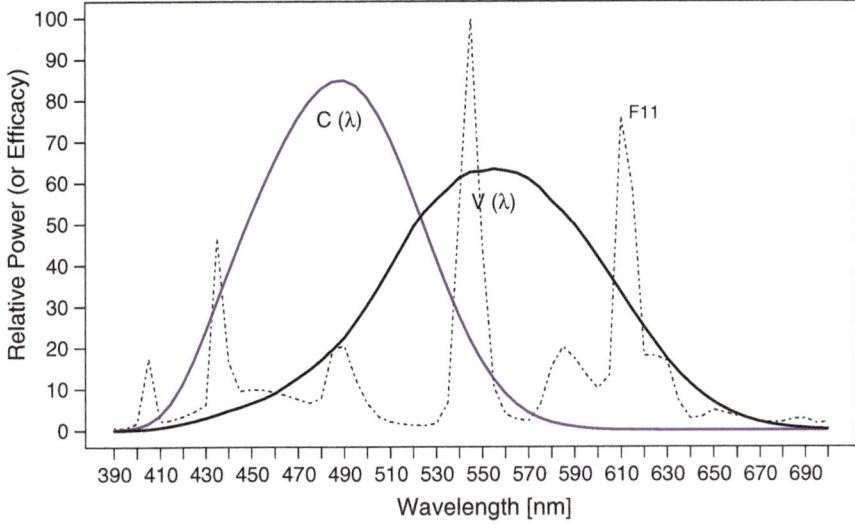

Fig. 2.13 Comparison of spectral response of the visual system (V-lambda) and the circadian system (C-lambda) to the spectral power distribution a narrow tri-band fluorescent lamp having a color temperature of 4000° K (the CIE illuminant F11)

Fig. 2.14 Mobile cart platform with laptop, HDR camera and CCD spectrometer

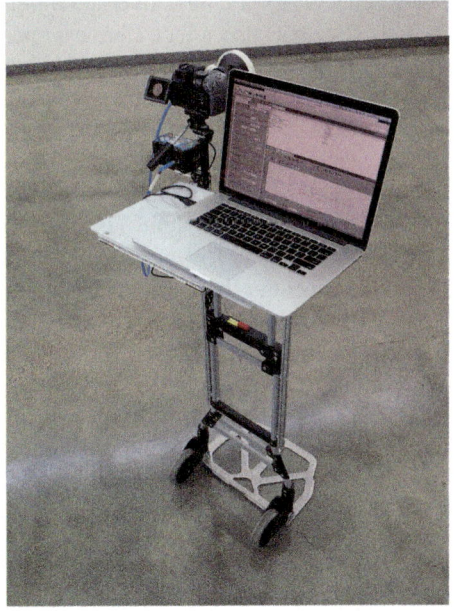

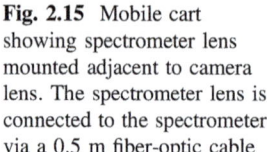

Fig. 2.15 Mobile cart
showing spectrometer lens
mounted adjacent to camera
lens. The spectrometer lens is
connected to the spectrometer
via a 0.5 m fiber-optic cable

circadian system to light can be found in Amundadottir et al. (2013). These lighting
related effects are of course overlaid on the myriad of physical and mental impacts
in daily life that also effect alertness, performance etc. over time so that clearly
disaggregating the lighting effects is a challenge. The framework developed by
Andersen et al. (2012) includes a schema to segment the day into three discrete
periods of analysis. These are, 6:00–10:00 AM (circadian resetting), 10:00–18:00
(alerting effects of daylight), and 18:00–6:00 (bright light avoidance, dim light
only). Access to bright, circadian effective light in the morning is most critical for
resetting the circadian system. Therefore, emerging CBDM metrics such as sDA or
UDI are problematic for the assessment of circadian potential of a space because
they do not account for the time during the day when a daylight stimulus is present.
Analysis should prioritize the interval from 6:00 to 10:00 AM. However, it is
important to note that exposure to bright light during the 10:00–18:00 period may
be desirable (and preferred) by occupants for its potential to improve alertness. The
task of developing novel daylight metrics and performance criteria specifically for
the evaluation of circadian entrainment in buildings is discussed in Sect. 2.4.3.

2.4.1   Daylighting for Circadian Entrainment

Daylight is an attractive alternative to electrical lighting for maintaining human
circadian entrainment indoors due to its spectrum (e.g. Fig. 2.12), intensity, general
availability, and potential to be introduced into spaces via windows and skylights.

Enabling designs that ensure the appropriate spectrum, timing, intensity, and duration of light to maintain healthy circadian entrainment will require a new set of performance objectives, measurement techniques, and assessment tools.

The first step is to address how light is measured. Due to the difference in the spectral response of the circadian system (C-lambda) from the visual system (V-lambda), the standard unit of illuminance (photopic lux), is problematic for quantifying the lighting conditions required to reset the human circadian system (Lockley et al. 2003). A number of efforts have emerged to rationalize how lighting outcomes can be determined and assessed in biologically meaningful terms. Researchers have proposed models of the spectral sensitivity of the circadian system that can be used to relate the SPDs from various light sources to a stimulus effect (e.g. nocturnal melatonin suppression (Rea et al. 2012), or perceived alertness (Andersen et al. 2012)). The model developed by Rea et al., which is applied to assess the circadian stimulus potential of the spaces shown in Figs. 2.16 and 2.18, is based on published studies of nocturnal melatonin suppression using lights of various SPDs. The model relates a given SPD to a circadian stimulus effect from 0 (0%) to 0.7 (70%) characterizing the relative effectiveness of the source as a stimulus, assuming a 1-hour exposure time. A publically available circadian stimulus calculator is provided to convert various light sources to units of circadian light (CL_A) and Circadian Stimulus (CS) for relative comparison of light source spectra (LRT 2016).

Table 2.1 presents the predicted circadian stimulus effect from various light sources using the model developed by Rea et al. (2012). The table can be used to determine the level of vertical illuminance (lux) at the cornea that must be achieved to produce circadian stimulus effects ranging from 10 to 70% for daylight (D65, clear sky with sun) and three common electrical light sources: LED 2700 K, 34-Watt T-12 linear fluorescent, and Halogen 3277 K. For example, to achieve a 20% circadian stimulus effect, an occupant must be exposed to 103 lx of daylight (D65) at the eye over a period of one hour. To achieve the equivalent stimulus effect with light from a 34-Watt T-12 "cool white" linear fluorescent lamp, the eye-level vertical illuminance must be increased by a factor of three, to 306 lx. A present, Figueiro et al. (2016) recommend exposure to a CS of 0.3 or greater at the eye for at least 1 h in the early part of the day (equivalent to 180 lx, D65).

Table 2.1 Circadian stimulus effect from various light sources

(CS) (%)	D55	D65	D75	LED 2700 K	34WT-12LF	Halo. 3277 K
10	66	46	40	86	131	59
20	146	103	89	190	306	131
30	255	180	156	337	530	231
40	423	301	261	568	870	390
50	730	523	455	1005	1470	690
60	1520	1110	970	2220	2950	1520
70	127,000	98,500	89,000	NA	NA	NA

Fig. 2.16 Workspace illuminated with fluorescent lighting

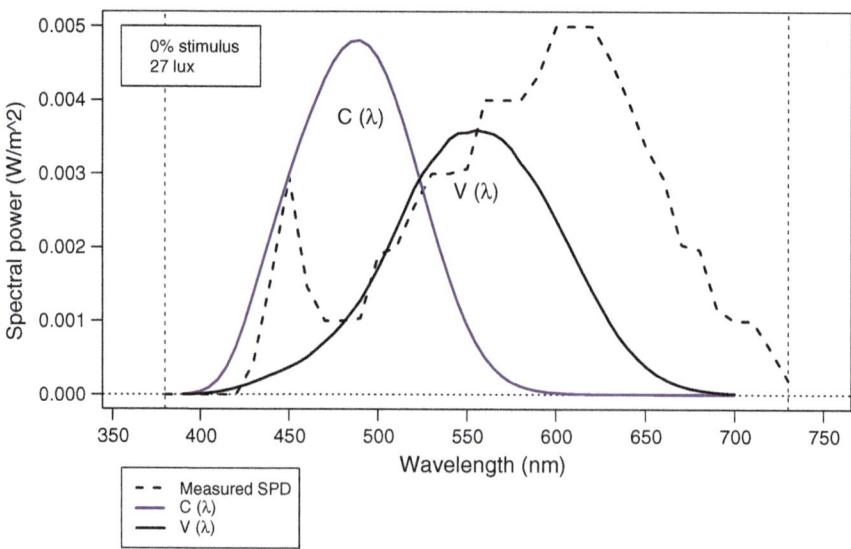

Fig. 2.17 Measured spectral power distribution, vertical illuminance (lux) and calculated circadian stimulus effect for Fig. 2.16 camera viewpoint

Alternatively, Lucas et al. (2014) have proposed a melanopic spectral efficiency function following the concept of melanopic illuminance introduced by al Enezi et al. (2011). Using a publically available calculator (Lucas et al. 2016), users can

Fig. 2.18 Workspace illuminated with daylight

calculate the resulting melanopic illuminance (lux) of various lighting conditions to understand and assess their biological impacts.

Researchers are also beginning to propose new approaches that seek to present a more holistic assessment of the effectiveness of a given lighting condition. For example, Rea and Bierman (2016) have proposed a universal luminous efficacy function (U-Lambda), which is proposed as a basis for setting light source efficacy requirements to serve multiple end user needs for light (e.g. color rendering, circadian regulation, scene brightness). Amundadottir et al. (2016) have proposed a unified framework to evaluate non-visual spectral effectiveness of light, which includes an online calculation and visualization tool (EPFL 2016) that can be used to compare the non-visual spectral effectiveness of various light spectra in terms of melatonin suppression, melatonin phase shift, and perceived alertness. Table 2.2 presents an example comparison of various common light source

Table 2.2 Biological impact of various light sources and photopic illuminances

Melatonin suppression (%)	EML	A (Lux)	F 11 (Lux)	D 65 (Lux)	LED 95 (Lux)
0.5	17	29	27	16	14
5	34	56	52	31	27
25	56	95	87	52	45
50	77	129	118	71	62
75	105	176	161	97	84
95	176	296	272	162	142
99.5	341	575	526	315	275

spectra (CIE A, CIE F11, CIE D65, and LED 95) in terms of Equivalent Melanopic Lux (EML), lux, and melatonin suppression (ranging from 0–99.5%). It should be noted that the framework developed by Amundadottir et al. incorporates a lens transmittance model to account for the relative loss in retinal exposure due to age of the observer. The outcomes presented in Table 2.2 are calculated assuming a 65-year-old observer.

2.4.2 Field-Based Measurement Practices

Because occupants are not well-equipped to report the circadian effectiveness of lighting conditions based on their own visual perception, and conventional photometric sensors are biased towards longer-wavelength light sources, new procedures are needed to measure and assess varying levels of circadian effectiveness, both during design and post-occupancy, where physical conditions may differ from design intent (e.g. due to window occlusion to control glare or increase visual privacy). And, the measurement condition must represent the conditions experienced by the human eye. This adjustment to conventional measurement practice creates several challenges, some of them obvious. First, the human eye is positioned vertically, requiring the measurement point to be oriented on a vertical, rather than horizontal plane. Second, occupants' viewpoints are likely to change over time, both regarding viewpoint location and view direction. Consequently, appropriate assumptions for the position and view direction of occupants are needed. Small changes to interior obstructions (e.g. partitions or furniture) can have large effects on levels of light reaching the eye. Consequently, it is a challenge to identify from what viewpoints in a space circadian effective lighting should be assessed, and what assumptions are most appropriate to account for potential obstructions. The challenge of view position is addressed in the examples presented in the following sections in context with additional parameters of light intensity, spectrum, duration and timing.

In the field, instrumentation capable of accurately measuring the SPD of light reaching the eye is needed to assess the relative effect of various light sources (and combined SPDs of multiple light sources) at various viewpoint locations in buildings. Figures 2.14 and 2.15 present one approach to address this need developed by the author, which uses a mobile cart platform to enable systematic evaluation of SPD in the field at adjustable eye-height levels (Burkhart and Konis 2016). The cart includes a digital Charge Coupled Device (CCD) spectrometer (model = OceanOptics JAZ-COMBO, effective range 300–750 nm, lens = cosine-corrected PTFE diffusing material) which is calibrated for measurements of absolute irradiance. The lens of the spectrometer is mounted adjacent to a High Dynamic Range (HDR) enabled CCD camera and connected to the spectrometer with a 0.5 m fiber-optic cable. The HDR camera enables point-in-time SPDs to be referenced to concurrent images acquired at near-identical viewpoints. These HDR images serve as a visual record of the scene and can be analyzed to evaluate glare and luminance conditions associated with SPD measurements.

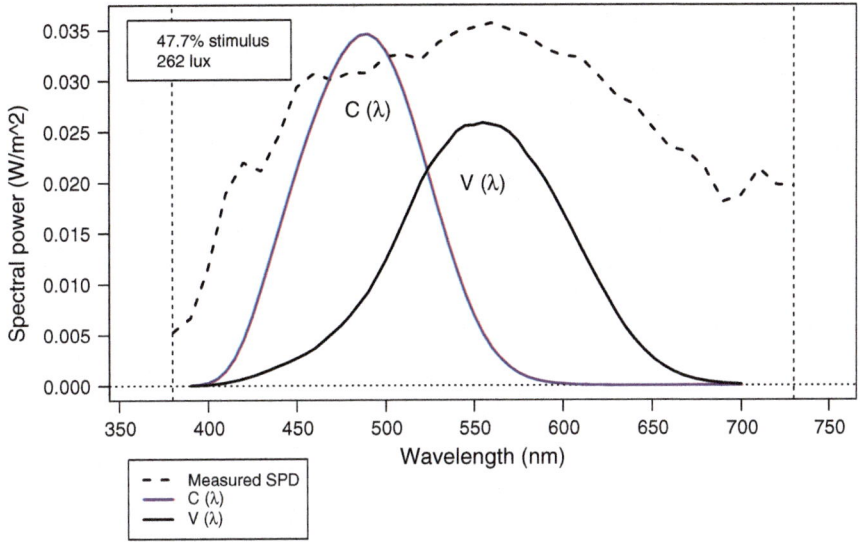

Fig. 2.19 Measured spectral power distribution, vertical illuminance (lux) and calculated circadian stimulus effect for Fig. 2.18 camera viewpoint

Figures 2.16 and 2.17 provide the outcome of a point-in-time evaluation using the mobile cart. Figure 2.16 shows a view of a work area illuminated exclusively by standard fluorescent lighting. The grey represents the luminous efficacy function (V-lambda) and the solid black curve indicates the response function of the circadian system (C-lambda). The measured global vertical illuminance at seated eye-level (27 lx) indicates that the light level is sufficient for photopic vision. However, note that the majority of the measured SPD falls outside the circadian response function (C-lambda) (Fig. 2.17). By applying the mathematical model developed by Rea et al. (2012) for quantifying circadian stimulus potential for a given SPD, which is based on a range from 0 to 70%, the lighting condition is found to be insufficient for circadian stimulus (0%). In contrast, Fig. 2.18 shows a similar work area illuminated exclusively with daylight. Despite the deployment of window shading devices on all windows, the measured global vertical illuminance at seated eye-level (439 lx) is higher, and the lighting condition is found to be sufficient to achieve a high level of circadian stimulus (55%) (Fig. 2.19).

2.4.3 Developing Circadian Daylight Metrics and Performance Criteria

There are currently no regulations governing lighting design to support circadian entrainment in buildings. Nor is there a consensus for the appropriate minimum light exposure threshold to ensure effective circadian stimulus, or for how long it must be present. Designers interested in addressing the need for daylight access for circadian entrainment are faced with a translational challenge. Knowledge of the biological effects of light is based on a limited body of data and work from disciplines of neuroscience and photobiology, where translation of research outcomes to design practices is not direct or often clear. However, there is a growing interest in the development of guidance and requirements for circadian lighting. One such example is The Well Building Institute's WELL Building Standard (IWBI 2016). In order to evaluate and refine the performance of a given design, available scientific findings must be examined to establish criteria for the appropriate timing, intensity, duration, and spectrum of light required for effective circadian entrainment. Additionally, assumptions must be made for the patterns of occupancy and even the view directions of occupants in each space. The Well Building Standard includes a Circadian Lighting Design precondition (option 1) which implements a minimum threshold of 250 EML (equivalent to 226 lx from D65), assessed vertically at eye-level, which must be available for at least four hours each day and can be provided at any point during the day. While the current version of the WELL circadian lighting pre-condition is problematic in that it does not specify the time period during the day when an effective stimulus must be present, and overlooks the challenges and assumptions needed for assessments of light exposure at the eye, it represents an important first step in efforts to translate available scientific knowledge into performance requirements to better ensure that buildings effectively support the health and well-being of occupants. It anticipated that the specific requirements and criteria, and their underlying assumptions, will be revisited as the relationships between spectral distribution, duration, timing, and intensity of light exposure for optimal circadian health are further clarified by the research community.

Theoretical knowledge and scientific findings are now sufficient to explore how architectural designs can serve to orchestrate effective patterns of daylight for circadian entrainment. Questions remain for how to appropriately evaluate design outcomes. Recently, Inanici et al. (2015) developed a simulation procedure to more accurately compute the spectral content of light for the purpose of analysis using circadian lighting indicators such as EML or CS. The procedure is currently implemented in a software tool (Grasshopper plugin) entitled "Lark Spectral Lighting[3]" which can be used by designers to analyze luminance renderings and irradiance data to obtain point-in-time calculations of EML or CS. Yet, even with the capability to accurately simulate the spectral content of light for a given viewpoint, there is still the task of appropriately interpreting, summarizing and visualizing simulation outcomes to inform the design process. To address this need, a novel area-based circadian daylight metric for building design and evaluation has been developed by the author (Konis 2016), which can be used to assess and differentiate the performance of various daylighting strategies during the design phases of a project, or to examine existing spaces based on the frequency with which an effective circadian stymulus is present. An example application of the

---

[3]http://faculty.washington.edu/inanici/Lark/Lark_home_page.html.

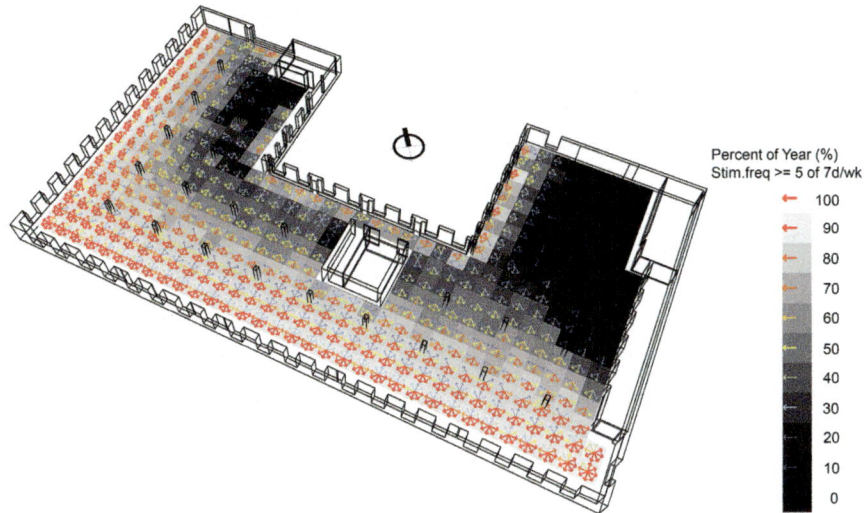

Fig. 2.20 Perspective view of building floor plate showing annual result for the percentage of analysis hours during the circadian resetting period (7:00–10:00 AM) where a minimum stimulus frequency of 71% (5 of 7 days/week) was achieved

metric is demonstrated in Fig. 2.20 through Fig. 2.23. Readers are encouraged to refer to the full paper (Konis 2016) for a detailed description of the metric and its calculation procedures.

Figure 2.20 shows the analysis result for a daylit office building floor plate located in downtown Los Angeles. A plan view is presented in Fig. 2.21. The geometry of the floor plate and fenestration is modeled after the location of the architectural design firm Perkins + Will's Los Angeles office. However, the example analysis presents the potential for daylighting prior to the addition of interior elements such as

Fig. 2.21 Plan view showing same result as previous figure

non-structural walls, workstation partitions, or interior furnishings. Procedures using annual, climate-based daylight modeling of eye-level light exposures are applied to map the space in regard to the availability of a circadian-effective daylight stimulus.

Because the biological effects of light exposure are not instantaneous, a novel indicator (referred to as "stimulus frequency") is applied to assess the frequency an effective stimulus is present over a window of time (e.g. 7-day period). While the minimum frequency needed to maintain healthy circadian stimulus is not known, it can be argued that measurement locations that have more frequent availability of an effective stimulus should be valued over those where availability is less frequent. The results in Fig. 2.20, reported for each view vector analyzed, show the percentage of the year where a stimulus frequency of at least 71% (5 of 7 days/week) is achieved. The daylighting potential of each location is then mapped based on the outcome of the best-performing vector (see Fig. 2.22). A stimulus is considered sufficient for a given day if a vertical light exposure of at least 250 EML is achieved throughout the portion of the circadian resetting period (7:00–10:00 AM) when the space is assumed to be occupied. Results can be used to identify and visually examine building zones where long-term occupancy may lead to disruption of the circadian system in the absence of supplemental electrical lighting capable of effective circadian stimulus.

Figure 2.23 presents an annual visualization of daylighting performance relative to varying levels (or grades) assigned to evaluate variations in levels of *entrainment quality*, where the "quality" of circadian entrainment is considered to diminish as the daily availability of an effective stimulus becomes less frequent over the moving 7-day analysis window. In Fig. 2.23, the percentage of analysis area falling into each *entrainment quality* grade category is reported on a scale ranging from 0 to 100% of the total analysis area. Designers can interpret Fig. 2.23 to understand seasonal changes in the spatial availability of a circadian stimulus as well as the

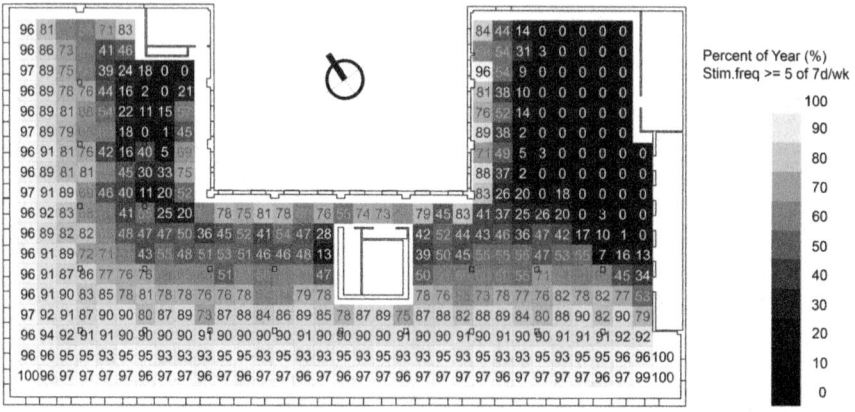

Fig. 2.22 Numerical mapping of the percentage of analysis hours during the circadian resetting period (7:00–10:00 AM) where a minimum stimulus frequency of 71% (5 of 7 days/week) was achieved

Fig. 2.23 Annual result showing daily variation in Circadian Effective Area (CEA) for each *entrainment quality* grade (*A, B, C, D* or *F*)

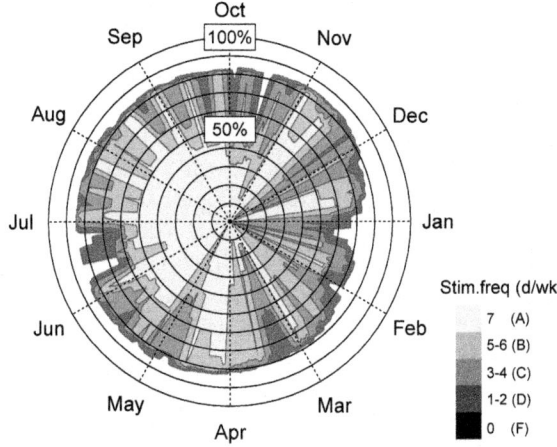

Table 2.3 Annual mean circadian effective area (0–100%) achieved for each *entrainment quality* grade

A	B	C	D	F
7d/wk	5-6d/wk	3-4d/wk	1-2d/wk	0d/wk
37.8	23.2	13.6	5.4	20

varying levels of *entrainment quality* achieved. For example, between the months of May to October, 40–50% of the analysis area achieves an *entrainment quality* grade of "A," indicating that a stimulus is present on a daily basis (i.e. 7 days within any 7-day period) within this area. Similarly, nearly 80% of the analysis area is shown to achieve some level of effective stimulus for most of the year, however the entrainment quality is often lower (e.g. only available on 2 of 7 days at some locations) and more variable. The annual Circadian Effective Area (CEA) falling into each *entrainment quality* grade category is summarized in Table 2.3 and can be used to make relative comparisons between various daylighting strategies during design. For example, the design objective would be to increase the percentage of analysis area falling into the higher-grade categories (e.g. A and B) and minimize area within the lower categories (e.g. C, D and F).

2.4.4   Limitations and Future Directions of Circadian Daylighting

Understanding how buildings orchestrate 24-hour patterns of light and dark is a critical frontier of research for assessing and rating the Indoor Environmental Quality (IEQ) of a number of common building types. Theoretical knowledge, expert judgment and emergent scientific findings are sufficient to begin to propose performance criteria that have the potential to be achieved through thoughtful architectural design. This section described parallel ongoing simulation and field-based efforts to examine the applicability of daylight the primary light source for circadian stimulus in buildings. The preliminary circadian daylighting metric (Konis, 2016) provides an important new capability to designers for quantifying and understanding the circadian potential of a given design as well as to identify biologically dark spaces in existing buildings, which require remediation or repurposing. The metric provides an additional objective for parametric simulation and optimization frameworks to rapidly explore and optimize the impact of a large combination of building parameters on the circadian potential of architectural space.

Unlike prior lighting and daylighting performance indicators, where applicability can be readily evaluated in the field by pairing physical measurements with occupant subjective assessments, the applicability of circadian daylight metrics for improving the health and well-being of occupants is much more complex, and will require novel methods to examine both short and long term health outcomes from daylighting design strategies in use. While these challenges are substantial, establishing feedback loops linking building design and occupant health outcomes is critical for improving quality of life in urban environments.

2.5   Visual Comfort

The balance of daylight transmission with the avoidance of glare is a central performance objective for effective daylighting. However, glare is rarely studied during the design process. This is largely due to the complexity of detecting and evaluating the dynamic patterns of luminance in daylight spaces and mapping how these patterns may affect the comfort and behavior of occupants. As noted in Sect. 2.2, maximum horizontal illuminance thresholds (e.g. 1000 lx), are currently implemented as proxy indicators for glare in CBDM metrics (e.g. UDI, sDA/ ASE). However, in modern work environments, visual tasks are often screen-based. With the visual task oriented vertically, direct view of the solar disc or extreme luminance contrasts between windows and indoor surfaces can often become sources of glare and are unlikely to correlate well (if at all) with measures of horizontal illuminance. As designers increasingly seek to improve access to daylight and window views for occupants, the ability to evaluate and address glare will be a critical factor in achieving effectively daylit spaces. This section discusses the potential and the limitations of existing and emerging approaches for evaluating glare.

2.5.1   Glare

Glare can generally be divided into three categories: (1) disability glare, (2) discomfort glare, and (3) veiling glare. Disability glare is defined as the disabling of the visual system to some extent by light scattering in the eye (Vos 1984) usually from very bright sources. Discomfort glare is defined by the IEA SHC Task 21 as:

"a sensation of annoyance caused by high or non-uniform distributions of brightness in the field of view." Alternatively, the Commission Internationale de l'Éclairage (CIE) defines discomfort glare as: "visual conditions in which there is excessive contrast or an inappropriate distribution of light sources that disturbs the observer or limits the ability to distinguish details and objects." Dynamic changes in lighting conditions that require rapid visual adaptation (e.g. from dark to light, or from light to dark) can also cause visual discomfort. Finally, veiling glare is the reduction in contrast of an image due to the reflection of a bright light source on the image, such as the reflection of bright windows on a computer monitor. Unlike disability glare, there is no well-understood mechanism for the cause of discomfort glare, although fluctuation in pupil size (Fry and King 1975) as well as distraction (Lynes 1977) have been suggested. Observation of daylit buildings in use often reveals the deployment of shading devices to address aspects of all three glare categories (e.g. Figure 2.24), which can in turn lead to significant reductions in daylight transmission, electrical lighting energy reduction, and visual connection to the exterior.

Figures 2.24, 2.25 and 2.27 present real examples of three common daylighting conditions that result in visual discomfort for building occupants. Lighting conditions were evaluated using a High Dynamic Range (HDR) enabled digital camera and software post-processing to produce calibrated luminance maps (Fig. 2.28) using a technique documented in (Konis 2012). This evaluation technique, and several of the most common metrics for glare analysis are discussed in detail in the following sections. Figure 2.24 shows a perimeter zone workstation where glare is caused by direct view of the solar disc. Despite the deployment of interior fabric roller shades, which supplement the additional solar control provided by an exterior perforated metal screen (50% openness) and solar control film (VLT = 0.23) applied to the facade glazing, the shade fabric openness factor of 0.03 (3%) is insufficient to completely block direct view of the solar disc, leading to luminances in excess of 50,000 cd/m^2 in the occupant's field of view.

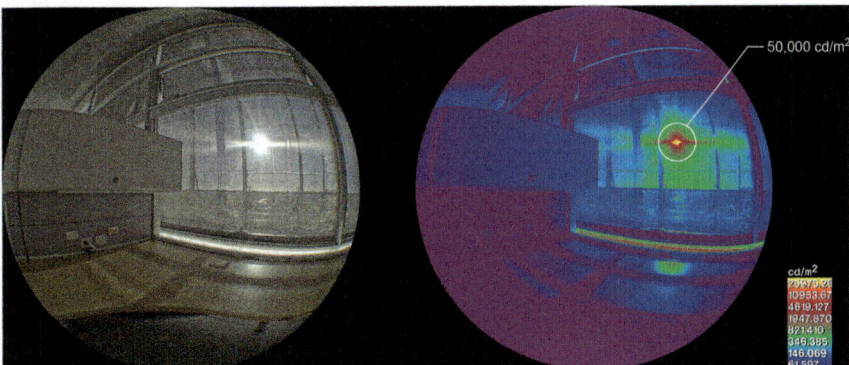

Fig. 2.24 Direct view of solar disc from perimeter zone workstation (*left*) and falsecolor luminance map (*right*)

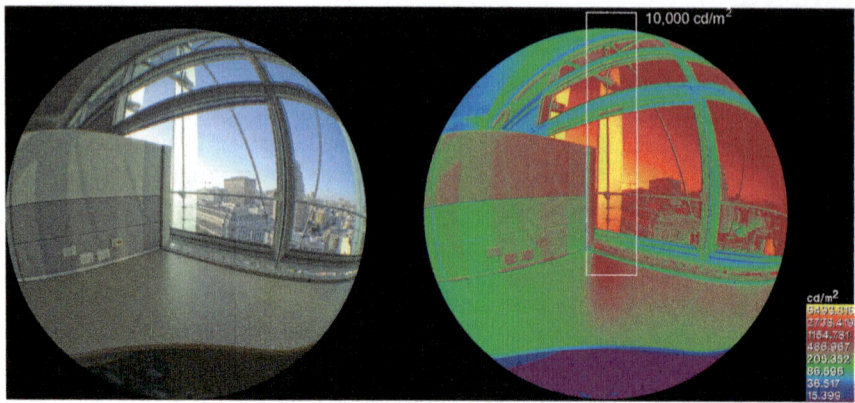

Fig. 2.25 View of exterior shading device surface in excess of 10,000 cd/m² from perimeter zone workspace

Fig. 2.26 Exterior view of translucent vertical louvers shown in Fig. 2.25

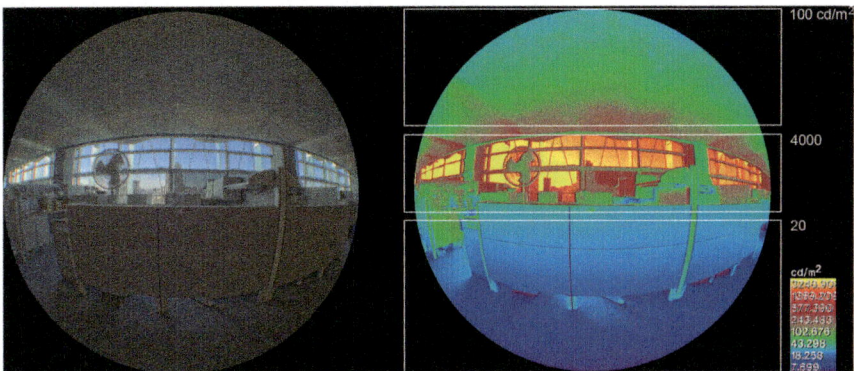

Fig. 2.27 View of facade glazing from core-zone of open-plan workspace in large daylit office building (*left*) and falsecolor luminance map highlighting contrast in luminance between facade glazing and interior surfaces

Fig. 2.28 Field installation of High Dynamic Range (HDR) enabled camera for acquisition of time-series measurements

Figure 2.25 shows an example where discomfort glare is produced from direct sun that is intercepted and diffused through the translucent louvers of an exterior shading system (Fig. 2.26). The louvers, when in direct sun, both reflect sunlight onto perimeter zone workstations (causing distracting luminance contrasts) as well as transmit diffuse light causing the entire louver surface to exceed luminance levels of 10,000 cd/m^2 on a daily basis. In comparison to Fig. 2.24, where visual discomfort was caused simply by the absolute magnitude of the glare source luminance, visual discomfort in Fig. 2.25 is caused by the excessive luminance contrast between the exterior fin surface (10,000 cd/m^2) and the interior surfaces in the field of view ($\sim$200 cd/m^2), which result in a ratio of over 50:1.

The example presented in Fig. 2.27 shows the view of facade glazing from a viewpoint in the core-zone of open-plan workspace in large daylit office building. In this example, the contrast in luminance between facade glazing (4000 cd/m^2) and interior surfaces (20–100 cd/m^2) exceeds a ratio of 40:1 and is likely to be a source of visual discomfort.

2.5.2 Daylight Glare Metrics

Concurrent with the reemerging interest in the daylighting of buildings in the 1960s, a study was conducted by (Hopkinson and Bradley 1960), to develop a metric to evaluate glare from large area sources (e.g. windows). The experimental setup consisted of a large illuminated diffusing screen (the light from the closely packed fluorescent lamps was diffused by an opal plastic screen), which provided a uniform luminance condition. The source size was varied from a small point source (10–3 sr) to the whole field of view, and the source luminance was varied between 3.5 and 15,500 cd/m^2. Subjects reported their subjective impressions of glare on a scale ranging from "just perceptible" to "just intolerable." The perception of glare depended not only on the brightness of the source but also on the size of the source as seen by the viewer, the viewers position relative to the source, and the surrounding scene luminance. The Daylight Glare Index (DGI) was derived and correlated to these subjective impressions.

$$DGI = 10\log 0.478 \sum_{i=1}^{n} \frac{L_s^{1.6} \cdot \Omega^{0.8}}{L_b + 0.07 \cdot \omega^{0.5} \cdot L_s}$$

L_s source Luminance (cd/m^2)
L_b background Luminance (cd/m^2)
Ω solid angular subtense of source modified for the effect of the observer in relation to the source (sr)
ω solid angular subtense of source at the eye of the observer (sr).

Equation 1. The Daylight Glare Index (DGI).

The DGI can be applied to predict the level of visual discomfort from windows by providing values for the parameters identified above (Equation 1). The DGI was recommended by the International Energy Agency (IEA) Solar Heating and Cooling (SHC) Program Task 21 daylighting performance monitoring procedures (IEA 2000) as the appropriate metric for predicting visual discomfort in daylight spaces. However, a number of other glare metrics have been proposed for use in evaluating visual discomfort from windows. These include: (1) the Unified Glare Rating (UGR) (Einhorn 1998), recommended by the Commission Internationale de l'Eclairage (CIE) and the ASHRAE Performance Measurement Protocols (PMP) for commercial buildings (ASHRAE 2010), and (2) the CIE Glare Index (CGI) (Einhorn 1969, 1979).

Until the last 10 years all complex glare metrics involve variations of the same basic relationship between the four parameters of glare source luminance, solid angle subtended by the glare source, the angular displacement of the source from the observer's line of sight, and the general field of luminance (i.e. "background" luminance) (Equation 2).

$$Glare = \int \left(\frac{L_S^{a_1} \cdot \omega_S^{a_2}}{L_b^{a_3} \cdot P^{a_4}} \right)$$

L_s source Luminance (cd/m^2)
ω_s solid angle of source
L_b background Luminance/adaptation luminance
P Position index.

Equation 2. Relationship of the four parameters of glare used in complex glare metrics.

New research after 2000 resulted in the Daylight Glare Probability (DGP) (Weinhold and Christoffersen). The DGP (Equation 3) describes the fraction of disturbed persons, caused by glare from daylight and is reported over a range from 0 to 1, with three semantic thresholds: "imperceptible," "perceptible," and "disturbing" glare, corresponding to DGP values of (0.35, 0.40, and 0.45) respectively. The DGP equation was developed from statistical analysis on a dataset of human-factors assessments collected in daylight test facilities (full scale office mock-ups) at two locations (Copenhagen and Freiburg) with more than 70 subjects. In contrast to other complex glare formulae, the DGP equation adds a term for Vertical Eye illuminance (Ev), which was found to improve the correlation of the model with users' responses.

$$DGP = c_1 \cdot E_v + c_2 \cdot \log \left(1 + \sum_i \frac{L_{s,i}^2 \cdot \omega_{s,i}}{E_v^{a_1} \cdot P_i^2} \right) + c_3$$

E_v vertical Eye illuminance (lux)
L_s source Luminance (cd/m^2)
ω_s solid angle of source
P Position index

$$c_1 = 5.87 \cdot 10^{-5}$$

$$c_2 = 9.18 \cdot 10^{-2}$$

$$c_3 = 0.16$$

$$a_1 = 1.87$$

Equation 3. The Daylight Glare Probability (DGP) equation.

Recent research (Suk et al. 2013, 2016) has also explored simplified calculation methods aimed at providing clearer guidance for designers by identifying the basic elements of potential glare in a scene (absolute luminance and contrast ratio). Suk et al. define these as Relative Glare Factor (RGF) and the Absolute Glare Factor (AGF). Values obtained for each factor can be considered by designers to understand the dominant glare factor as well as predict the level of perceived discomfort through comparison to threshold values proposed by the researchers based on human-factors studies. Researchers have also begun to explore the application of multiple glare metrics in a multiple regression model and found that models combining multiple metrics predicted subjective visual discomfort better than a single metric alone (Van Den Wymelenberg 2012; Jakubiec et al. 2016).

2.5.3 Application of Glare Metrics Using HDR Images

In contrast to the relatively small, uniform, and stationary glare sources with constant brightness produced by electric lighting, the glare sources produced by windows vary in brightness, are constantly changing in size and position, and are usually distributed non-uniformly across a large area (e.g. a window or facade). Visual comfort calculations depend not only on the locations and brightness of light sources, but also on the apparent size of the light sources as seen from a particular viewpoint (Ward 1992). This presents a difficult measurement problem to researchers using conventional photometric instruments (e.g. masked illuminance sensors, or spot luminance meters) because the observer's entire field of view must be sampled in order to capture the luminance, position, and size of the glare source

(s) produced by the sky conditions. In addition, due to the non-uniform lighting distributions common in daylit spaces, the boundary of the glare source is more difficult to define. High Dynamic Range (HDR) images, by acquiring scene luminance data on a "per-pixel" scale, provide the ability to record the size, position and luminance of an arbitrary number of potential glare sources in the field of view, potentially enabling greater accuracy in the detection of dynamic glare sources.

Figure 2.29 presents the same glare examples presented in Figs. 2.24, 2.25, and 2.27 evaluated with the analysis program *evalglare*. Evalglare is a software program based on the studies of Weinold and Christoffersen (2006) and was developed to detect and evaluate glare sources within a 180° hemispherical image given in the Radiance image format (.pic or .hdr). Evalglare reports the DGP for the given scene in addition to a number of other common glare metrics and includes a number of input parameters that can be manipulated to adjust the predicted outcome. The most significant input assumption is the specified threshold factor for glare source detection that can be a constant value (e.g. all regions that exceed 1000 cd/m^2), or a multiple of the average visual task luminance (e.g. all regions that exceed seven times the average luminance of a user-specified visual task area), or a multiplier of the average luminance of the entire scene (if no task view is given). The program operates with a default assumption that all regions that exceed 5 times the visual task (or entire scene) should be treated as a glare source. Figure 2.29 compares the original .hdr image (left) with the check file produced by Evalglare (right), using the "cut" field of view according to Guth, which presents the total field of human vision as limited by facial structure.

2.5.4   Dynamic Glare Evaluation

While a single "point-in-time" evaluation of glare may be valuable for static lighting environments, it offers limited feedback on the success or failure of a given daylighting design over daily and seasonal changes in sun and sky conditions. Understanding visual comfort performance requires assessing the time-varying patterns of luminance from specific views, including the effect of active shading use, and making assumptions for how daily and annual patterns impact occupant acceptance and behavior. As a simple example, Fig. 2.30 presents a daylit scene from a real building over the course of 12 h in one day under predominantly clear sky conditions. Figures 2.31 and 2.32 show the calculated DGI and DGP outcomes at 5-minute intervals derived from HDR images acquired on site. Notably, the DGI and DGP daily profiles vary in their prediction of the severity of glare, with the DGP predicting glare exceeding the "disturbing" semantic threshold in the morning and afternoon (Fig. 2.31) and the DGI predicting a level of glare above "just acceptable," but below "just uncomfortable" (Fig. 2.32). While the glare metric predictions vary considerably throughout the day, it is unlikely that occupant comfort and acceptance change at the same rate, or correlate directly with "point-in-time" predictions. It is far more likely that occupants will form opinions about the visual comfort of their environment over a much longer time period, and

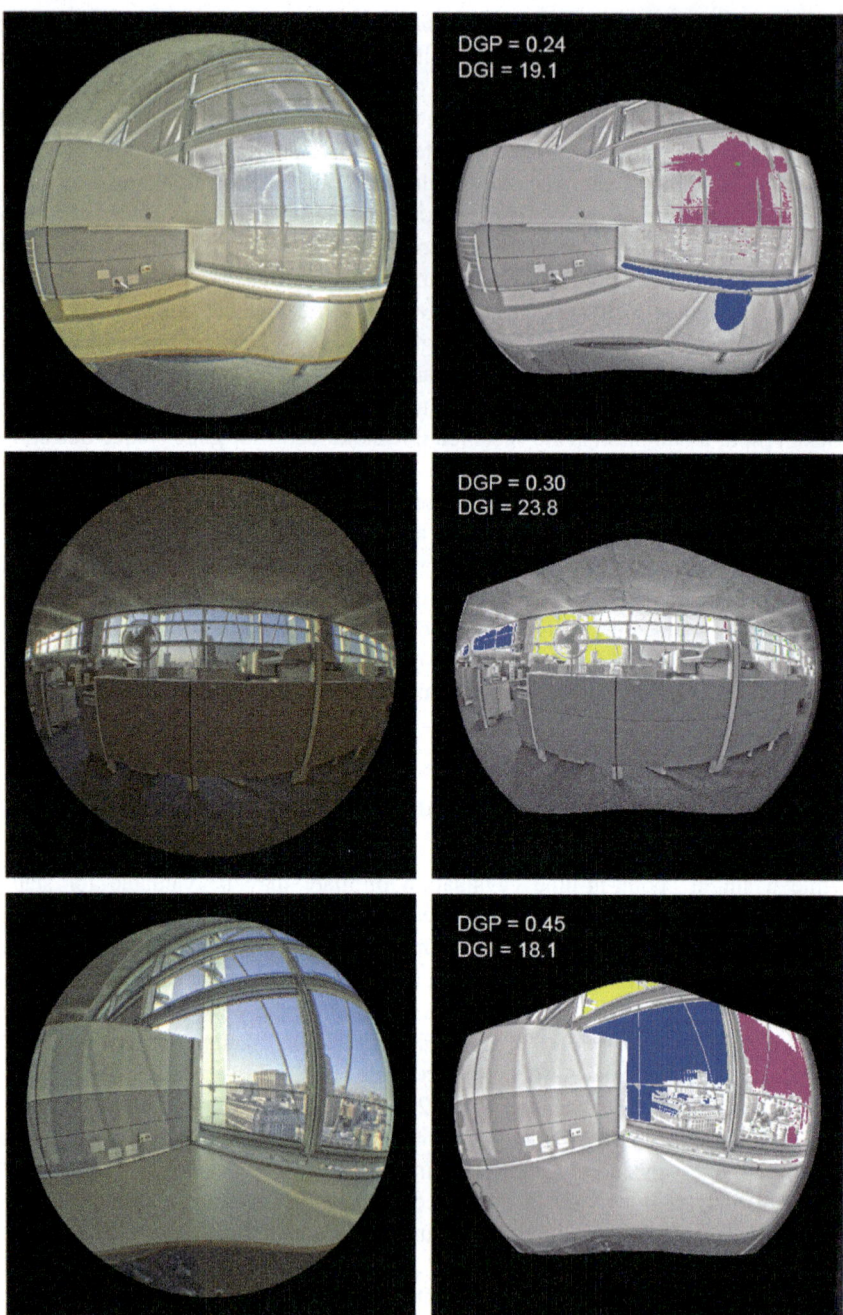

Fig. 2.29 Example field of view modification, glare source detection, and glare prediction (DGP and DGI) performed by the *evalglare* software tool on HDR images of real daylit scenes. Arbitrary colors are used to identify the glare sources detected

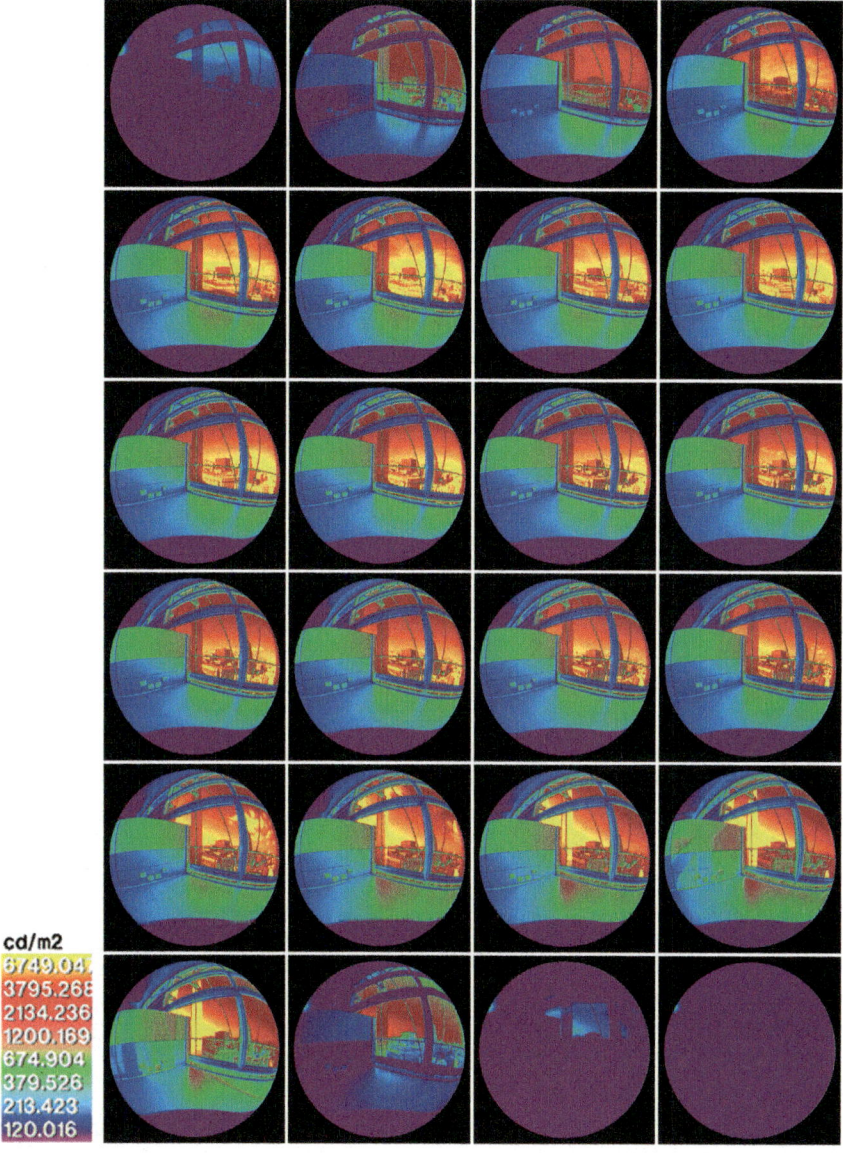

Fig. 2.30 Time-series representation of daily luminance pattern for window facing view from Fig. 2.25. An image is shown at every 0.5 h from 6:30 AM to 6:00 PM Standard Time October 25, clear sky conditions

will both adapt to, and modify their environment to reduce visual discomfort. The most common adaptation is to turn ones head away from the worst glare source and to lower available shading devices, which, if manually operated, are rarely retracted when the source of glare is no longer present.

Fig. 2.31 Time-series DGP prediction based on *evalglare* analysis of 5-minute interval HDR image data

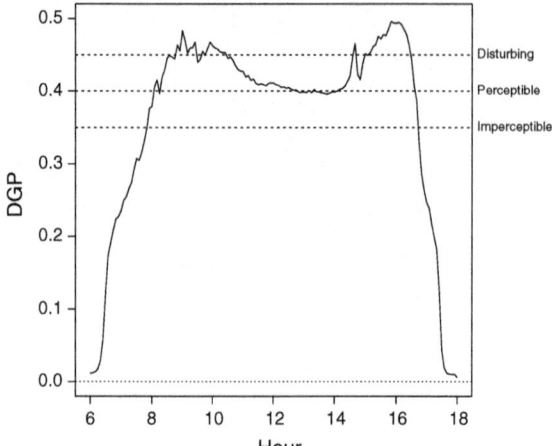

Fig. 2.32 Time-series DGI prediction based on *evalglare* analysis of 5-minute interval HDR image data

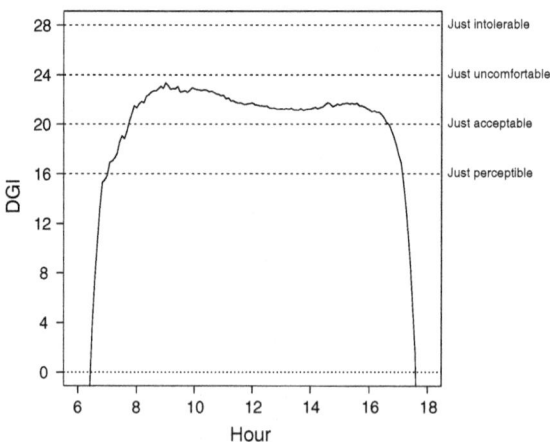

2.5.5 Frequency and Magnitude of Glare

In an effort to examine the frequency and magnitude of glare over various time periods (e.g. week, month, year), Weinold developed Dynamic Daylight Glare Evaluation (DDGE) (Weinold 2009). The approach applies the evalglare tool to time-series sets of images from a given viewpoint within a space. Time-series results for a specified period (e.g. annual, occupied hours) are then ordered by magnitude and examined relative to various proposed daylight glare "comfort classes." Class A, B, and C are used as a basis to differentiate performance outcomes as shown in Table 2.4. To visually examine the daily and seasonal occurrence of varying levels of glare for a particular viewpoint, Jakubiec integrated

Table 2.4 Daylight glare comfort classes defined by Weinold (2009)

	A	B	C
	Best class	Good class	Reasonable class
	95% of office-time glare weaker than "imperceptible'	95% of office-time glare weaker than "perceptible"	95% of office-time glare weaker than "disturbing"
DGP limit	<= 0.35	<= 0.40	<= 0.45
Average DGP limit within 5% band	0.38	0.42	0.53

Both limits (DGP, and average DGP within 5% band) must be fulfilled

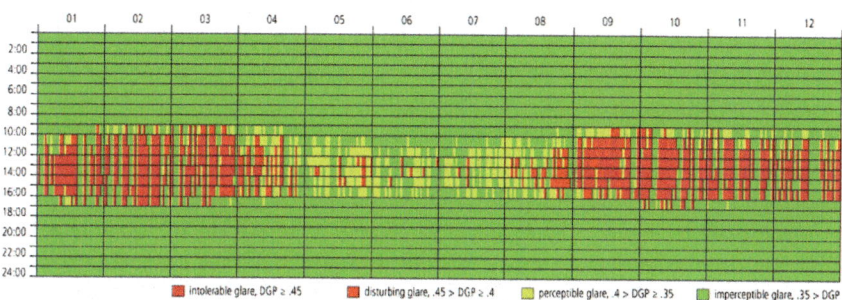

Fig. 2.33 Annual Daylight Glare Probability (DGP) simulation. The x-axis corresponds to 365 days of the year, the y-axis corresponds to time of day. *Red* and *orange* fields correspond to hours with intolerable or disturbing glare, respectively, *yellow* to perceptible glare and *green* to imperceptible. *Image credit* Alstan Jakubiec

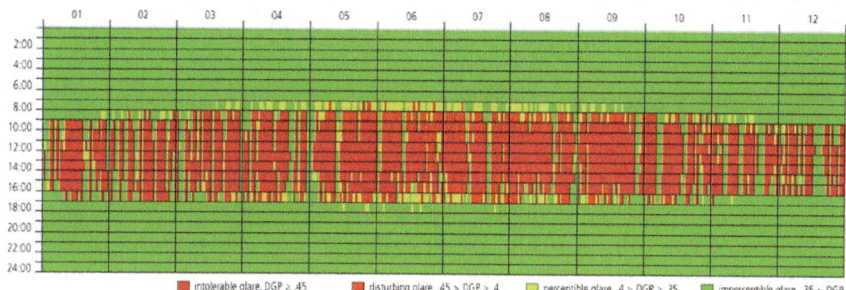

Fig. 2.34 Annual Daylight Glare Probability (DGP) simulation result for window-facing view orientation. *Image credit* Alstan Jakubiec

DDGE into the software DIVA-for-Rhino (Jakubiec and Reinhart 2012), to generate annual glare maps. Figure 2.33 presents an example of an annual outcome for a task-facing view. For comparison, Fig. 2.34 presents the annual result for the same location, but with a window-facing view.

2.5.6 *View-Direction Dependent Glare Evaluation*

To predict glare discomfort in open-plan office environments, it is necessary to evaluate all significant views in regularly occupied spaces within a project. This requires that dynamic glare evaluation include multiple view positions and, due to the ability of occupants to adjust their view direction, a range of view vectors from each position. To address this latter challenge, Jakubiec and Reinhart (2011) developed a concept called the "adaptive zone" and a simulation-based approach where cylindrical images (a 180° vertical, 360° horizontal view) overlaid with a view direction-dependent glare evaluation are used to predict levels of discomfort glare for a user-specified range of view orientations. Individual images can be composited into animations that can be used by designers to visualize the directionality of glare for a specific location (and range of view directions) of interest within a project. Figure 2.35 shows an individual "point in time" cylindrical representation of view and corresponding glare predictions for various available view directions. Jakubiec and Reinhart found that by applying the adaptive zone concept to a sidelit office with manually operated venetian blinds it was possible to "reduce the predicted hours of intolerable discomfort glare from 735 to 18 occupied hours per year and increases the annual mean daylight availability from 40 to 72%". Figures 2.36, 2.37 and 2.38 show view-direction dependent glare evaluations for Gund Hall, (Harvard Graduate School of Architecture), at various times during the year. The bars across the bottom of each image illustrate predicted levels of discomfort glare in the indicated orientation for each analyzed metric (green = imperceptible, yellow = perceptible, orange = disturbing, red = intolerable).

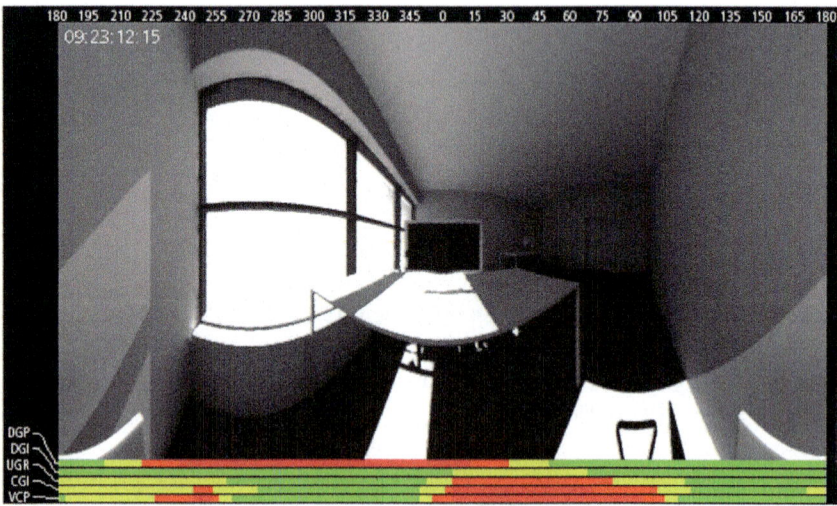

Fig. 2.35 View-direction dependent glare evaluations on September 23 at 12:15 PM in sidelit office space. *Image credit* Alstan Jakubiec

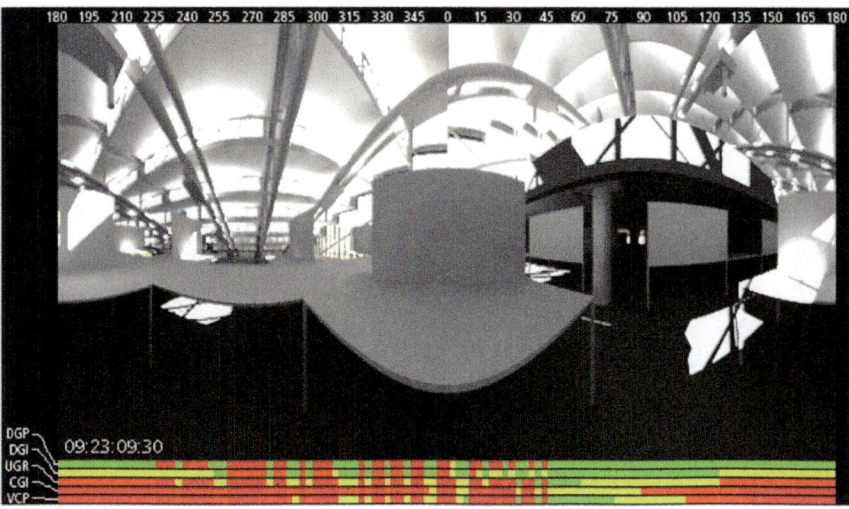

Fig. 2.36 View-direction dependent glare evaluations on September 23 at 9:30 AM in a large open-plan work space. *Image credit* Alstan Jakubiec

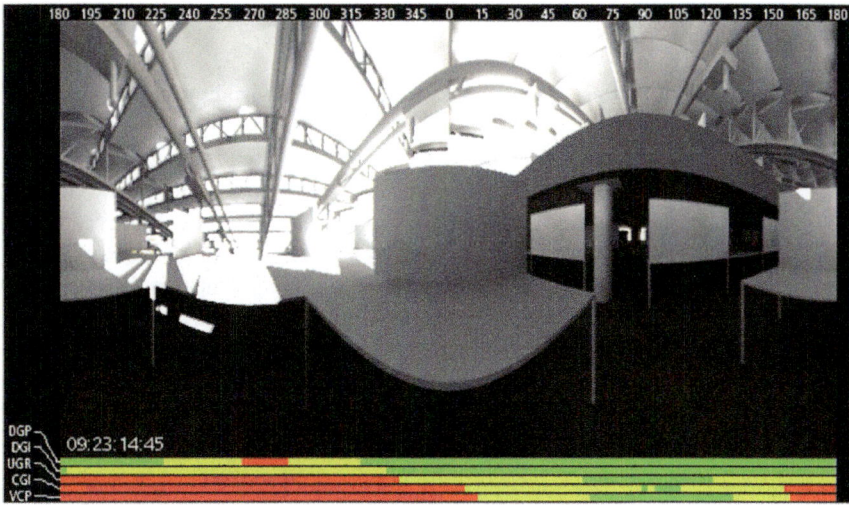

Fig. 2.37 View-direction dependent glare evaluations on September 23 at 14:45 AM in a large open-plan work space. *Image credit* Alstan Jakubiec

Fig. 2.38 View-direction dependent glare evaluations on December 21 at 12:45 AM in a large open-plan workspace. *Image credit* Alstan Jakubiec

2.5.7   Limitations and Future Directions of Visual Comfort Evaluation

The development of better methods and tools to predict visual discomfort in daylit spaces remains an active research topic. Currently there is no widely agreed-upon method to accurately predict discomfort glare in daylit environments. And while a single "point-in-time" evaluation of glare may be valuable for static lighting environments, it offers limited feedback on the success or failure of a given design over daily and seasonal changes in sun and sky conditions. Understanding visual comfort performance requires assessing the time-varying patterns of luminance from specific view positions and making assumptions for how hourly, daily and seasonal patterns impact occupant behavior and shade use. While quantitative, simulation-based methods have been developed, assumptions relating annual exposures to occupant outcomes are largely derived from very limited (and much shorter-term) laboratory-based occupant studies or on expert judgment rather than on extensive field validation. Further inquiry is needed to evaluate how occupants adjust shading and make other behavioral modifications in daylighted spaces, as well as how they form long-term opinions of visual comfort in dynamic daylit environments that include varying levels of glare, and how occupants prefer to manage trade-offs between levels of glare and other IEQ factors, such as access to a window view or higher daylight levels. For example, research by Tuaycharoen and Tragenza (2007) indicates that the absolute tolerance of glare from windows is related to the visual content of the view through the window, where higher pre-dicted DGI values will be tolerated for views rated positively. This supports studies from the 1960s when glare ratings based on electric lighting were first being adapted for use with daylight. One significant challenge in the application of computing annual climate-based daylighting metrics, modeling of occupant behavior and luminance-based glare analysis is in the development of equally complex Post Occupancy Evaluation (POE) mechanisms capable of validating the large collection of assumptions embedded in annualized performance outcomes.

Rather than working to establish a consensus for the most effective glare metric, behavioral model, or equation for annualizing the results of hourly daylighting simulations, designers may forgo a universal design paradigm based on a theoretical "standard observer" and begin to apply existing metrics and analysis tools to develop personalized, data-driven comfort models, drawing on the increasing availability of sensor feedback from real daylit spaces in use. For example, luminance maps acquired from low-cost HDR imaging devices (e.g. LBNL SkyCam), paired with contextual, behavioral and subjective data, can be analyzed on embedded computers to determine unique, real time comfort models. With enough data and time, these models might generate algorithms based on measured data that inform selection and optimization of the major design parameters. These models, in turn, can be shared within the design profession to improve understanding of user experience in buildings, as well as to inform the operation of dynamic glare control systems in the buildings themselves.

It is important to keep simulation outcomes in context with a holistic set of design options. For example, completely blocking the solar disc at the facade with 3-dimensional exterior screen may achieve the same visual comfort outcome as a simple, thoughtfully designed adjustable shade integrated into a workstation partition. The latter option provides occupants with personal task-level control over a thoughtfully considered dynamic range of lighting conditions, while maintaining views and transmission of sufficient ambient daylight to meet IEQ and energy objectives.

2.6 Visual Connection to the Outdoors

Greater emphasis on the provision of access to window views for all occupants is helping to invert conventional practices for the space planning of office buildings, placing open-plan offices along the perimeter of the floor plate and locating enclosed cellular office space in the core. For larger buildings, view requirements for the majority of regularly occupied space necessitate a transition from relatively "fat" floor plate buildings with a low surface-to-volume ratio to "thinner" more elongated building forms, with a higher ratio of surface-to-volume and often a more complex form. Finally, in addition to encouraging thinner floor plates, the adoption of emerging metrics aimed at quantifying and rating available views, such as the "view factors" now being adopted by voluntary rating systems like LEED, are a further incentive for designers to apply floor-to-ceiling facade glazing in order to achieve compliance for deep floor plate buildings, creating significant technical challenges for managing thermal and visual comfort along the perimeter.

Interest in the provision of views for all occupants is driven by a large body of research in the field of environmental psychology that supports the conventional wisdom that the provision of windows is an essential component of occupant performance, health, and well-being. In an effort to characterize these benefits, Collins (1975) conducted a review of available literature and reported windows serve a number of psychological functions, including view, stimulation, and the perception of spaciousness in addition to the provision of sunlight and daylight which were both shown to be desired by building occupants. Collins additionally reported that the absence of windows in spaces that were confined or static could result in adverse reactions from occupants. Later research in windowless workspaces by Heerwagen and Orians (1986) showed that occupants frequently decorate a windowless office with posters of outdoor scenes as a means of creating a "surrogate" window. Figure 2.39 presents an example of a "surrogate window" (right) installed in a medical office building. A staff member installed the "surrogate window" on the back surface of a sign directly in front of her field of view (left). The "surrogate window" is a detailed photograph of a large redwood tree surrounded by a forest landscape. What is notable about this example is that the view position is approximately 12 m from the facade and includes a large window view of an adjacent building, which delivers significant levels of daylight. An informal interview of the staff member revealed that "surrogate window" was installed due to the perceived poor quality of view content provided by the window. This individual example supports the theory presented decades ago by MC Lam in his seminal work, *Perception and Lighting as Formgivers for Architecture* (Lam 1977), where

Fig. 2.39 Example of a "surrogate window" (*right*) installed in a medical office building. A staff member installed the "surrogate window" on the back surface of a sign directly in front of their field of view (*left*)

Table 2.5 Biological needs for environmental information, after Lam (1977)

Location	With regard to water, heat, food, sunlight, escape routes, destinations, etc
Time	And environmental conditions which relate to our innate biological needs
Weather	As it relates to the need for clothing and heating or cooling, the need for shelter, opportunities to bas in the beneficial rays of the sun, etc
Enclosure	The safety of the structure, the location and nature of environmental controls, protection from cold, heat, rain, etc
The presence of other living things	Plants, animals, and people
Territory	Its boundaries and the means available within a given environment for the personalization of space
Opportunities for relaxation and stimulation	Of the mind, body, and senses
Places of refuge	Shelter in time of perceived danger

he outlines a list of important biological needs for environmental information (Table 2.5), which go far beyond the provision of view. If these needs are not well served by designers, occupants will make modifications to the extent possible to better serve these various needs.

In addition to the availability of a view, the content of the view is shown to have an effect on psychological well-being. The most consistent finding is the preference for natural over built views (Farley and Veitch 2001). Windows with natural views were found to enhance work and well-being in a number of ways including increasing job satisfaction, interest value of the job, perceptions of self-productivity, perceptions of physical working conditions, life satisfaction, and decreasing intention to quit and the recovery time of surgical patients (Farley and Veitch 2001). The view of a natural scene through a window (either real or simulated) has also been proposed as a means of reliving stress (Kaplan 1993; Ulrich 1991). The content of the view can also affect the preference of occupants towards the size and shape of the window, with relatively smaller windows being acceptable for distant views and larger windows required for views of nearby objects (N'eman and Hopkinson 1970). Studies have also shown that access to a window view can have a measurable relationship to changes in office worker performance. In a field-based investigation conducted in two large office buildings in California, the Heschong Mahone Group reported that better access to a window view was found to consistently predict better performance (CEC 2003).

Access to a distant view has also been linked to eye health. In modern office environments where workers spend increasing amounts of time viewing computer screens or workstation partitions, the distant view provided by windows allows changes in eye focus distance to give the eye muscles a chance to relax. Because the focus distance required for ocular muscles to relax is significantly greater than the dimensions of most buildings, a window view of distant scenery provides an important alternative focus for the eyes.

Given the body of research on the importance of window view for occupant health and well-being, the provision of a satisfactory level of visual connection to the outdoors through window views is a critical performance objective. A number of parameters can be considered evaluating view. These can be separated into factors considering the availability, amount, and quality of visual connection to the outdoors. Each parameter is discussed in the following sections.

2.6.1   Window Size and Aperture Configuration

Many European and Scandinavian building standards include provisions for view, which are often incorporated with daylighting requirements. One such example, first published in 1935, is the German Standard on daylighting (DIN 5034, "Daylight in Interiors"). Part 1 of DIN 5034 specifies minimum window sizes based on room size as well as requirements for the configuration of the window aperture. According to DIN 5034-1 (2011), the top edge of visually transparent window glazing must be a minimum of 2.2 m (7.21 ft) above the finished floor height, and the bottom edge cannot exceed 0.9 m (2.95 ft). In addition, the sum of window widths must meet or exceed 55% of the room width, leading to a minimum window-to-wall ratio requirement of approximately 30%.

While provision of a window view is not a requirement for office buildings in the U.S., the desire to specify view requirements in green building rating systems has led to a need to define measurable criteria for window views. In the current version of LEED (v4) (USGBC 2016), the concept of a "view factor" is introduced, based on a study of office worker performance and the indoor environment (CEC 2003). Calculations of the view factor result in a numerical score from 0 to 5 determined by the smaller of the lateral and vertical view angles for a specified viewpoint. As defined in the report: "*A view rating of 5 almost completely filled the visual field of the observer seated at the cubicle. A view of 4 filled about one-half of the visual field. A view of 3 represented about one-half the size of a view 4, but still with a coherent view. A view rating of 2 represented a narrow and typically fractured view. A view rating of 1 represented a glimpse of sky or sliver of the outside environment.*" Table 2.6 provides minimum view angles for each view factor score. Compliance with the view factor option in the current version of LEED (v4) requires a view factor of 3 or greater. Figure 2.40 shows the lateral and vertical view angles achieved for a viewpoint located 3 m (9.8 ft) from the facade, which results in a view factor of 4. While the original view factor scores were determined from observational studies including moveable furnishings and other obstructions common in office spaces after occupancy, the LEED calculation procedure allows for non-permanent obstructions to be excluded (Fig 2.41).

Table 2.6 View factors

View Factor	View angle	
	Min-max (°)	Gray zone range (°)
1	1–4	
1 or 2		4–5
2	5–9	
2 or 3		9–11
3	11–15	
3 or 4		15–20
4	20–40	
4 or 5		40–50
5	50-90	

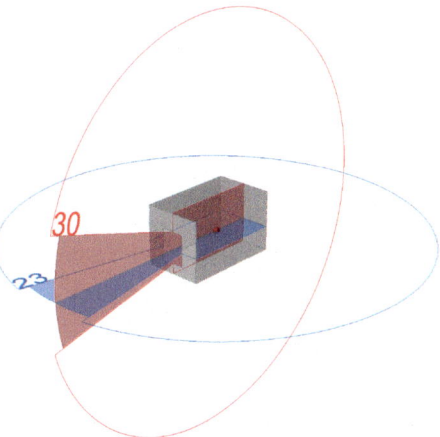

Fig. 2.40 Lateral and vertical view angles achieved for a viewpoint located 3 m (9.8 ft) from the facade

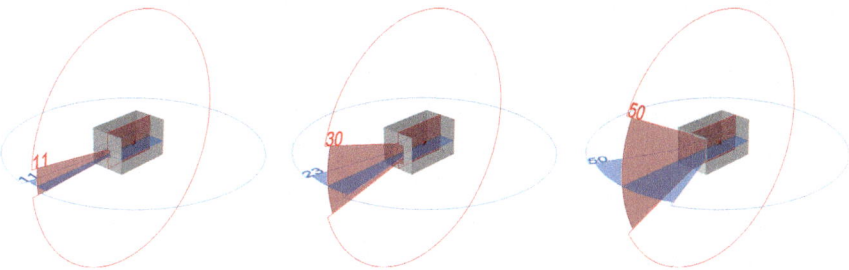

Fig. 2.41 View factors of 3,4,5 for a seated view-point 3 m from the facade

2.6.2   Distance of Occupants from Windows

The distance of occupants from windows is another important parameter for assessing visual connection to the outdoors, and has significant implications for building form. For example, DIN 5034 (2011) requires that all workspaces must be located within 10 m of a window. This limit restricts the floor plate depth of German office buildings, leading to relatively "thinner" forms than their U.S. counterparts, and more frequent use of courtyard and atria formal arrangements due to the greater ratio of skin to volume. While not an explicit distance limit, the LEED requirement to provide, "unobstructed views located within the distance of three times the head height of the vision glazing," leads to a similar distance limit of approximately 10 m from windows for typical finished floor-to-ceiling heights (e.g. 3 m). The Nordea Bank Building (Fig. 2.42), designed by Henning Larsen Architects, presents a contrast to typical large commercial office building planning. A primary objective of the building form is to provide the best opportunities for all of Nordea's employees to work in an environment connected with daily and seasonal changes in daylight and views to the outdoors. Atriums are placed in the center of the building mass and serve to spatially connect the first floor (level 01) to the upper floor (level 07) creating a feeling of unity between the various work zones within the large project. The open place offices are arranged along the exterior of the floor plates adjacent to the facade, providing a direct visual connection to the exterior environment for all regularly-occupied work areas (see Chap. 5 for a more detailed description of the project).

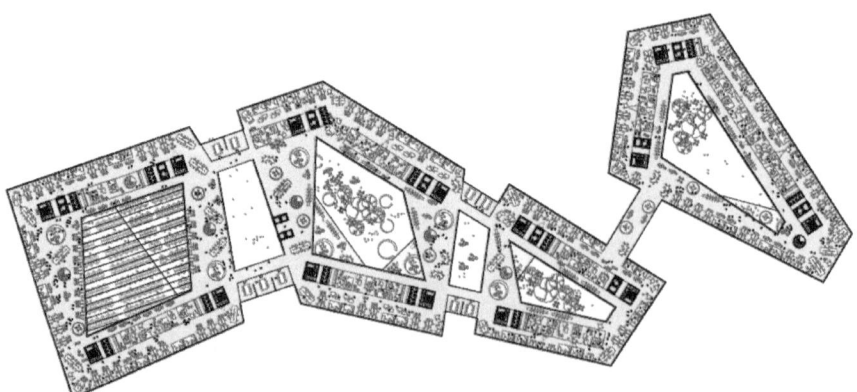

Fig. 2.42 Nordea Bank Headquarters typical *upper* level floor plan. *Image credit* Henning Larsen Architects

2.6.3   Provision of Multiple Views

The number, direction, and aggregate view angle of available window views can also be used to evaluate the view potential from a given location. Providing multiple views can enable greater awareness of exterior phenomena (e.g. changes in weather, activities) as well as provide more diverse visual content (e.g. both urban and natural views). Perhaps the most practical benefit of views from multiple directions is the possibility of preserving an unshaded window view when other views require shading for solar and glare control. The quantity of views can be evaluated using a number of indicators including (1) the total number of distinct window views, (2) the total visual angle of available window views, as well as the distribution of views over the occupant's horizontal field of view. For example, the LEED v4 compliance option requires, "multiple lines of sight to vision glazing in different directions at least 90° apart." The available number of views should be considered in context of the occupant's primary visual task view, if known. For example, the views available from the occupant's primary task view (while seated) may be valued higher than the views available when standing and/or looking away from the primary visual task. Figure 2.43 shows an example analysis for one test point located at seated eye-height on an open-office floor plate. The analysis uses a view rose technique to visualize the total number of window views, the horizontal view angle of each view, and the total horizontal view angle (119 of 360°) by

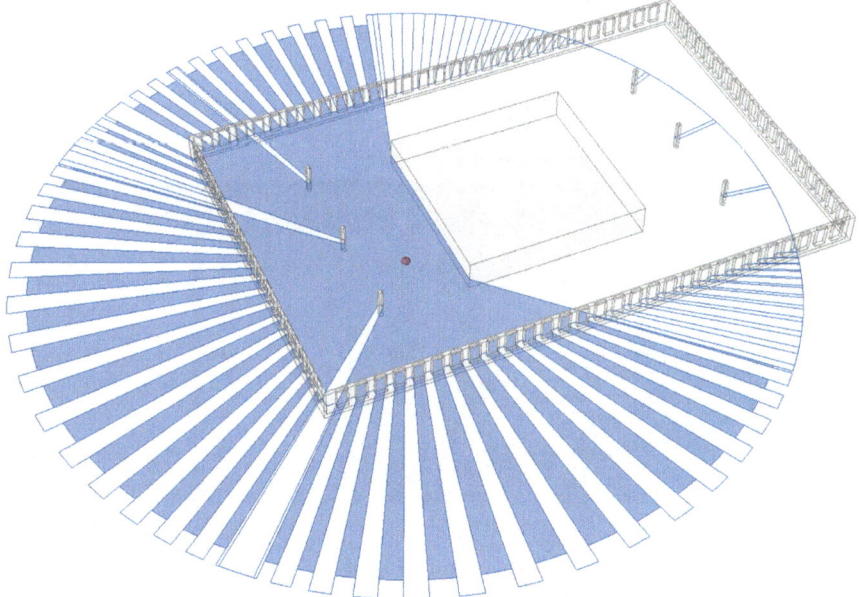

Fig. 2.43 Example line-of-sight analysis for a specified workstation location in an open-plan office floor plate. Analysis was performed using the "view rose" component in Ladybug

determining the direct lines of sight reaching a specified distance from the test point
(e.g. 40 m).

2.6.4 View Content

Attributes of view content can also be used to evaluate view quality. For example,
Fig. 2.44 compares a view of a traditional Chinese garden (The Garden of Flowing
Fragrance, Liu Fang Yuan[4] (left), with a view of blank wall opposite a narrow
daylit void space (right). The view of the garden includes a number of key attributes
that contribute to a quality view. These include view of (1) flora and fauna, (2) the
sky, and (3) movement (e.g. surface of water, branches and leaves of trees), (4) the
presence of people, and (5) a distant view. These attributes, in addition to many
others, help to enable a complex emotional process defined by the eminent biologist
Edward O. Wilson as *Biophilia*: "the innately emotional affiliation of human beings
to other living organisms" (Wilson 1986, p. 31). Recognizing that attributes such as
these are undervalued in conventional design practices relative to their importance
for maintaining human psychological well-being, scholars have worked to develop
and identify *biophilic design* practices (e.g. Kellert and Heerwagen 2008) as well as
include requirements for view content into green building compliance criteria. For
example, the current version of LEED requires that, "views that include at least two
of the following: (1) flora, fauna, or sky; (2) movement; and (3) objects at least 25
feet from the exterior of the glazing." While designers can rarely construct natural
landscape settings, designers can survey the visual assets available for each project
using the attributes of biophilic design as a filter to prioritize the organization and
orientation of program space and building form.

Fig. 2.44 Window view to a high-quality view content (traditional Chinese garden), (*left*).
Window-view to low-quality view content (adjacent blank wall), (*right*)

---

[4]http://www.huntington.org/chinesegarden/.

2.6.5   *Visual Transparency and Openness Factor*

For fenestration systems with interior or exterior solar and glare control elements that screen or partially occlude the window view, the openness factor is an additional parameter with significant impacts on view quality. Control of excessive solar heat gains is one of the primary challenges for low energy daylit office buildings. While designers can easily reduce window size and add coatings or solar control films to reduce solar gains, contemporary designers rarely take this approach due to the negative impacts on daylight availability and views. Instead, designers are increasingly using exterior solar control screens over large areas of facade glazing to reduce solar loads while creating larger window views that preserve screened or partially-occluded views for occupants. Figure 2.45 shows an example of the perforated metal screen used for solar control on the southeast-facing facade of the San Francisco Federal Building. The screen is composed of small, regularly-spaced circular perforations which achieve a 50% openness factor at normal incidence (Fig. 2.46, right). While Fig. 2.45 (right) demonstrates a lack of visual transparency to the interior from outside the building, the views from inside the building adjacent to the facade are largely preserved (Fig. 2.46, left), despite the physical occlusion of over half of the view.

Figure 2.47 shows the automated exterior solar control screens applied as a facade retrofit (revitalisierung) to the Haupthaus KfW building in Frankfurt. In contrast to the previous example, the Haupthaus screens are composed of a glazed sandwich panel with an interlayer of expanded metal. Compared with the previous example, the expanded metal screen results in a significantly lower openness factor at normal incidence (Fig. 2.48, left). However, significant visual information is

Fig. 2.45 Perforated metal screen used for solar control on the southeast-facing facade of the San Francisco Federal Building. *Note* that the openness factor (50% or 0.5) assumes a view at normal incidence (perpendicular) to the plane of the facade. The *left* image shows how the apparent transparency of the material diminishes significantly for oblique views. The *right* image shows how the exterior screen completely blocks views of the building interior during daylight hours

Fig. 2.46 Interior views looking through the facade glazing and exterior perforated metal screen at distances of 1 m (*left*) and 0.1 m (*right*)

Fig. 2.47 Automated exterior solar control screens applied as a facade retrofit (revitalisierung) to the Haupthaus KfW building in Frankfurt (*left*) and interior view with screens deployed (*right*)

Fig. 2.48 The Haupthaus exterior solar control screens are composed of a glazed sandwich panel with an interlayer of expanded metal configured to completely block direct view of the solar disc from the interior while preserving a partial view to the exterior. Images are taken from the building interior at varying distances from the screen (0.2, 1, and 3 m)

preserved, with the content of the window view becoming clearer as the distance of the viewer from the screen increases (Fig. 2.48). In addition, the angular tilt of the expanded metal allows for increasingly open views in a downward direction, enabling views below the horizon to be preserved while increasingly blocking views to the sky that may include the solar disk. Finally, the panels can be completely retracted to enable unobstructed views when solar or glare control is not required.

2.6.6 Visual Clarity

In addition to openness factor, the clarity of the window view is an important design consideration for view quality. For example, DIN 5034 makes explicit provisions for the clarity of the view: "For this reason it is necessary to provide windows with transparent, undistorted and neutrally colored glazing at the eye level of persons standing or sitting in a room." Similarly, LEED (v4) requires that, "view glazing in the contributing area must provide a clear image of the exterior, not obstructed by frits, fibers, patterned glazing, or added tints that distort color balance." (U.S.G.B.C. 2015). Distortion of the view can result from the application of frit patterns (as shown in Fig. 2.49), prismatic glazing, optical light-redirecting films, or simple light diffusing polymer materials. Tinting or coloration of the view results from alternation of the spectral content of light due to changes made to the chemical formulation of glass to improve solar control and typically produce neutral grey, bronze and blue-green colors.

Fig. 2.49 Horizontal frit pattern applied to portions of the glazed facade of New York Times building

2.6.7 Limitations and Future Directions Related to View

While the parameters outlined in this section present a useful means of evaluating the amount and quality of views during design, it is important to note several limitations with current approaches. First, the view angles calculated during design may omit the presence of interior objects such as furniture and partitions that block direct line of site for occupants when seated. Second, calculation of direct line-of-sight views and view angles discounts the significant impact of shading devices that are often deployed to address issues related to glare and solar overheating near windows. Figure 2.50 presents an example from the San Francisco Federal Building, contrasting the view content available (left) with the views preserved through the south east facade following the retrofit application of manually operated interior roller shades (openness = 0.03) and a solar control film to address issues related to discomfort glare and occupant solar overheating. This example illustrates the difficulty in preserving quality visual connection to the outdoors, particularly from core zone workstations, without taking an integrated approach to the design of the facade, dynamic shading systems and controls, and the workstations themselves. Third, view factor calculations do not take into consideration the content of the window view, and thus may overestimate the benefit of increased window area near the floor or ceiling that may add little additional visual information of value to occupants. Fourth, quality window views require effective glare control. Therefore, designers may overestimate the value of views that include the path of the sun but do not completely block occupant views of the solar disc, or views with a high level of luminance contrast between the window view and adjacent interior surfaces.

As fenestration systems become more optically complex, the most effective method of differentiating view quality during design will likely be through full-scale physical mockups and human observational studies. Full-scale test facilities such as the LBNL Flexlab (Chap. 4) present an ideal setting for such human factors evaluations, and provide the capability of evaluating human factors outcomes alongside energy and controls optimization objectives. Where physical observation is not practical, such as in the earliest stages of design, simulation

Fig. 2.50 View potential of southeast facing facade (*left*) and actual view from workstation located approximately 10 m (33 ft) from the facade

techniques incorporating image-based lighting (Chap. 3) can serve as a preliminary means of examining simulated views using actual visual context and scene luminances captured from the project site. Finally, as dynamic solar and glare control layers become more common, dynamic view-based metrics will be needed to appropriately differentiate systems based on the fraction of time during occupied hours when quality views are maintained.

Multiple parameters for evaluating the availability, amount, and quality of window views have been presented, including minimum window size, view factor, distance from windows, view content, view occlusion, and view clarity. However, it remains unclear how occupants relatively value trade-offs among these various parameters. To improve the fidelity of design assumptions and occupant satisfaction, it is important to examine buildings in use to assess the applicability of current view-based performance criteria as well as learn how occupants rank the importance of various performance indicators. Similarly, it is important to examine how occupants modify available views to address factors such as privacy and view, visual discomfort and solar control. These issues are discussed in detail in Chap. 6.

2.7 Solar Control and Thermal Comfort

In addition to controlling solar (shortwave) radiation indoors to minimize glare and space cooling loads, exposure to sunlight has a significant impact on occupant thermal comfort.

Because thermal comfort standards were developed assuming occupants would not be directly exposed to shortwave radiation, relevant standards such as ASHRAE Standard 55 (ASHRAE 2004) or ISO Standard 7730 (ISO 2015) do not account for the impacts of shortwave gain on the body of the occupant. Solar radiation falling directly on occupants creates additional, often substantial, thermal stress that is often beyond the capacity of cooling systems to offset. And, because the occurrence of direct sun varies spatially and temporally, systems that attempt to cool sunlit areas often cause thermal discomfort due to overcooling adjacent (non-sunlit) areas. As designers increasingly seek to achieve both daylit and thermally comfortable, energy efficient buildings, the standard design condition for occupant thermal comfort no longer resembles the internal and tightly controlled thermal zones in which existing thermal comfort standards are derived. The critical design condition for assessing thermal comfort in daylit buildings is the daylit perimeter zone (e.g. Figure 2.51), where, until recently, there have been no design tools available to study the effects of solar radiation on indoor thermal comfort.

Figure 2.51 presents an example of direct sun in an unoccupied south-facing perimeter zone workstation on the southeast facade of the San Francisco Federal Building. The image is representative of the original design intent for creating a thermally comfortable daylit perimeter zone through the application of spectrally selective facade glazing (SHGC 0.37) and an exterior solar control screen, with an openness factor of 0.5 at normal incidence. The combined effect of these solar

Fig. 2.51 Unoccupied south-facing perimeter zone workstation on the south-east facade of the San Francisco Federal Building

control layers leads to over an 80% reduction in solar transmission through the facade, and was considered acceptable by the design team for occupant thermal comfort as well as for the level of solar control needed to avoid supplementing the low-energy cooling strategy with an air-conditioning system (McConahey et al. 2002). However, as noted in a post occupancy evaluation of the building, (Konis 2012), the original design was subsequently retrofit with interior shades and a solar control film (solar energy transmission = 0.33) to address issues of occupant solar overheating and visual discomfort.

Recently, Arens et al. (2015) developed *SolarCal*, a model for predicting the effect of indoor solar exposure on occupant thermal comfort. The SolarCal model, "computes an increase in Mean Radiant Temperature (MRT) equivalent to short-wave gains from direct, diffuse and indoor-reflected radiation on a person" (Arens et al., 2015). The solar-adjusted MRT can then be used to compute the Predicted Mean Vote (PMV) using the ASHRAE-55 prescribed method to obtain a more realistic prediction of occupant thermal comfort in spaces with direct sun (Hoyt et al. 2014). Built on the formulae developed for SolarCal, Mackey (2015) developed a solar-adjusted thermal comfort "virtual manikin" integrated within the Ladybug/Honeybee (Sadeghipour 2013) suite of environmental analysis plug-ins for the 3D modeling software Rhinoceros. The thermal manikin software enables designers to compute the thermal sensation that is being experienced by occupants near windows and generate more accurate prediction of thermal comfort.

Figure 2.52 shows the solar adjusted radiant temperature across the surfaces of a thermal comfort manikin during the fall equinox (11:00–12:00). Radiant temperatures on surfaces of the body that exceed typical zone temperatures (e.g. 21–23 °C) indicate the need for additional (often substantial) space cooling and increase the

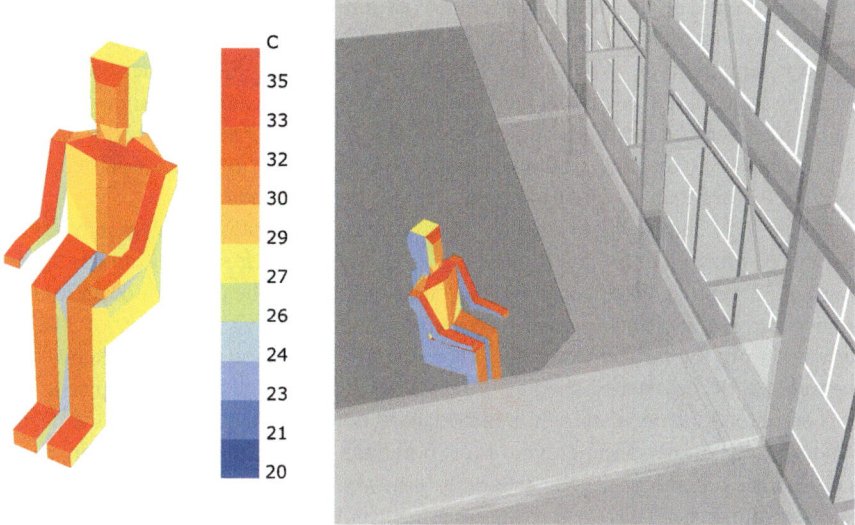

Fig. 2.52 Design condition solar adjusted radiant temperature on thermal comfort manikin (September 21, 11:00–12:00, clear sky conditions, no interior roller shades deployed on the facade)

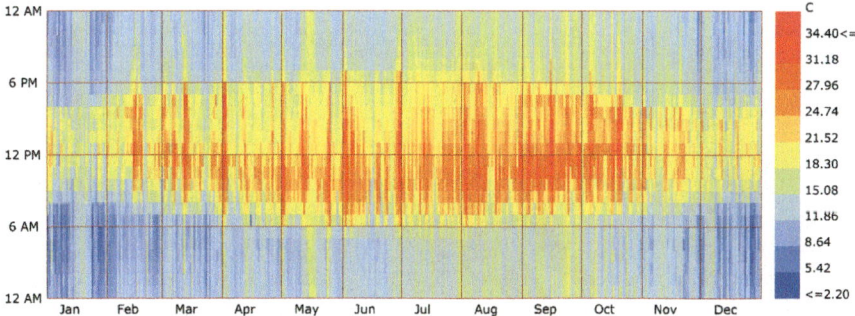

Fig. 2.53 Solar-adjusted MRT on an hourly basis throughout the year

MRT of the occupant relative to an equivalent zone without sun. In the case shown in Fig. 2.52, the resulting solar-adjusted MRT is 35 °C, and the MRT discounting the effects of solar radiation is 19.5 °C. Results can be produced on an hourly basis over an annual period to examine the frequency and magnitude of solar effects on MRT (Fig. 2.53).

Perhaps the greatest benefit for designers is in developing appropriate exterior shading strategies and in selecting material solar optical properties (e.g. glazing SHGC) in response to feedback on occupant solar-adjusted thermal comfort. While effective solar control is needed on an annual basis to avoid occupant modifications or more formal retrofits to the building facade, in early stages of design, a design condition representing critical solar control requirements can be used. For example, the hours of the year where peak solar-adjusted-MRT occur concurrent with peak outdoor temperatures. While the design team for the San Francisco Federal Building considered occupant thermal comfort in detail during design using the ASHRAE adaptive thermal comfort model (Haves et al. 2004), the resulting fraction of allowed solar transmission can be readily shown to produce thermal discomfort for occupants seated in sunlit areas of the perimeter.

Figure 2.54 shows the design condition solar adjusted radiant temperature on the thermal comfort manikin (September 21, 11:00–12:00, clear sky conditions) for the original (as-built) facade (letter B), the facade following the addition of an interior solar control film (letter C), and a hypothetical scenario with the exterior shading removed from the facade (letter A). Comparison of the various outcomes shows that the removal of the exterior shading would likely lead to extreme thermal discomfort for perimeter zone occupants. It is important to note that this (letter A) is the design condition for most commercial office buildings without external shading, and includes the reduction in solar heat gain provided by high-performance spectrally selective glazing (SHGC = 0.37). The retrofit outcome, (letter C), shows that the combined effect of three layers of solar control (exterior, glazing, and film), which achieve a combined SHCG of approximately 0.06 is sufficient to maintain occupant thermal comfort in the perimeter zone.

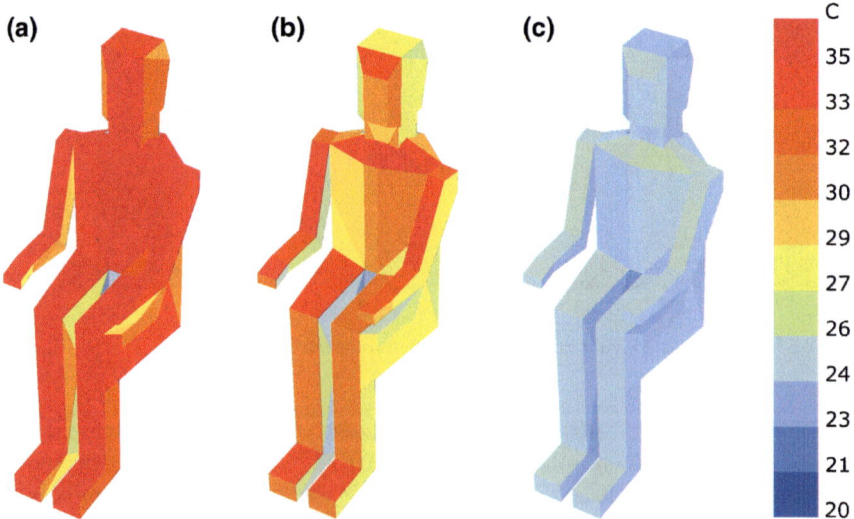

Fig. 2.54 Design condition solar adjusted radiant temperature on thermal comfort manikin for three cases: **a** Facade with exterior shading removed, **b** Facade as designed, **c** Facade with addition of solar control film retrofit (September 21, 11:00–12:00, clear sky conditions)

2.7.1   Limitations and Future Directions of Solar/Thermal Comfort Evaluation

Thoughtful consideration of the impact of shortwave solar radiation on occupant thermal comfort is critical in early stage design to establish a realistic baseline facade configuration for design development including the effects of shading. While dynamic facade systems (discussed in Chap. 3) present a technological approach for more effectively controlling solar exposures in the perimeter zone, it is important to note that the design and operation of automated systems require realistic assumptions for the range of thermal conditions acceptable to occupants. By enabling more realistic predictions of solar-adjusted thermal comfort on an hourly basis, the tools discussed above can be integrated with other annualized simulation approaches to serve as a basis for the operation of automated facade solar control systems that may be designed to dynamically modulate the allowable transmittance of fenestration. The final state of the dynamic facade thus must account for and prioritize the often contradictory requirements of cooling load control, daylight transmittance, glare control, thermal comfort management and view.

2.8   Conclusions

Efforts to achieve daylighting performance goals influence numerous building design parameters with impacts across a range of physical and temporal scales. These include project siting and orientation, form and massing, floor-plate depth, sizing and location of apertures, configuration of fenestration systems, zoning and sizing of mechanical HVAC and lighting systems, interior programming and furnishings, and many other parameters. Energy and occupant performance has one intrinsic time scale, impacts on occupant health and wellbeing may have a longer time frame. Will a building that performs well today also be a top performer in 10 or 20 years? Performance metrics, when integrated into the design process, help to enable a feedback loop to better understand how adjustments to individual parameters (and various combinations of multiple parameters), are likely to affect project performance over these scales. Through iteration, metrics can be used to go beyond compliance-based design outcomes to performance-based design processes that seek the optimum solution among multiple, (and sometimes conflicting) performance objectives. This latter task is dealt with in Chap. 4. Finally, measureable performance goals serve as a basis to compare the performance of the project in use with design intent to inform and refine future design efforts.

References

Amundadottir et al (2013) Modeling non-visual responses to light: unifying spectral sensitivity and temporal characteristics in a single model structure. In: Proceedings of CIE Centenary Conference "Towards a New Century of Light", Apr 15–16, 2013, Paris, France, p 101–110

Amundadottir ML, Lockley SW, Andersen M (2016) Unified framework to evaluate non-visual spectral effectiveness of light for human health. Lighting Res Technol 2016:1–24

Andersen M, Mardaljevic J, Lockley SW (2012) A framework for predicting the non-visual effects of daylight—part I: photobiology-based model. Lighting Res Technol 2012(44):37

Arens E, Hoyt T, Zhou X, Huang L, Zhang H, Schiavon S (2015) Modeling thecomfort effects of short-wave solar radiation indoors. Build Environ 88:3–9. doi:10.1016/j.buildenv.2014.09.004

ASHRAE (2004) ASHRAE. Standard 55-2004: thermal environmental conditions for humanoccupancy. American society of heating, refrigerating and air-conditioning engineers

ASHRAE *Performance Measurement Protocols for Commercial Buildings* (PMP) (2010) Atlanta: ASHRAE Inc. http://www.techstreet.com/ashrae/standards/performance-measurement-protocols-for-commercial-buildings?ashrae_auth_token=&gateway_code=ashrae&product_id=1703581

Bourgeois D, Reinhart CF, Ward G (2008) A standard daylight coefficient model for dynamic daylighting simulations. Building Res Info 36(1):68–82

Brainard GC, Hanifin JP, Greeson JM, Byrne B, Glickman G, Gerner E, Rollag MD (2001) Action spectrum for melatonin regulation in humans: evidence for a novel circadian photoreceptor. J Neurosci 21:6405–6412

Burkhart K, Konis K (2016) Daylighting evaluation of a leed platinum laboratory building: a post-occupancy study comparing performance in use to design intent. PLEA 2016 Los Angeles. Cities, Buildings, People: Towards Regenerative Environments, 11–13 July 2016

California Energy Commission (CEC) (2003) Windows and offices; A study of office worker performance and the indoor environment. Technical report no. 500-03-082-A-9. http://www.energy.ca.gov/2003publications/CEC-500-2003-082/CEC-500-2003-082-A-09.PDF

California Public Utilities Commission (CPUC) (2008) California long term energy efficiency strategic plan. http://www.cpuc.ca.gov/general.aspx?id=4125

Chang AMM, Scheer FA, Czeisler CA (2011) The human circadian system adapts to prior photic history. J Physiol 589(5):1095–1102

CIE (2004) Ocular lighting effects on human physiology and behavior. Commission Internationale de l'Eclairage Publication 158(2004):2004

Collins BL (1975) Windows and people: a literature survey. Psychological reaction to environments with and without windows. National Bureau Standards, Washington DC

DIN 5034 (2011) Tageslicht in Innenräumen, Deutsches Institut fur Normung

École Polytechnique Fédérale de Lausanne (2016) Interdisciplinary Laboratory of Performance-Integrated Design. SpeKtro, Interactive dashboard for exploring non-visual spectrum lighting.http://spektro.epfl.ch/ Accessed 24 July 2016

EIA (2012) Commercial building energy consumption survey (CBECS). Table 6. Electricity consumption by end use, 2012

Einhorn HD (1969) A new method for the assessment of discomfort glare. Lighting Res Technol 1 (4): 235–247

Einhorn HD (1979) Discomfort glare: a formula to bridge differences. Lighting Res Technol 11(2): 90–94

Einhorn HD (1998) Unified Glare Rating (UGR): merits and application to multiplesources. Lighting Res Technol 30(2):89–93

Enezi et al (2011) A "melanopic" spectral efficiency function predicts the sensitivity of melanopsin photoreceptors to polychromatic lights. J Biol Rhythms 2011(26):314

Farley KMJ, Veitch JA (2001) A room with a view: a review of the effects of windows on work and well-being. National Research Council of Canada (NRCC) technical report IRC-RR-136. http://nparc.cisti-icist.nrc-cnrc.gc.ca/eng/view/object/?id=ca18fccf-3ac9-4190-92d9-dc6cbbca7a98

Figueiro et al (2016) Designing with circadian stimulus. lighting design and applications (LD + A). The magazine of the Illuminating Engineering Society of North America (IESNA). Published October 2016

Fry GA, King VM (1975) The pupillary response and discomfort glare. J Illuminating Eng Soc 4 (4) http://dx.doi.org/10.1080/00994480.1975.10748533

Gooley JJ, Lu J, Chou TC, Scammell TE, Saper CB (2001) Melanopsin in cells of origin of the retinohypothalamic tract. Nat Neurosci 4(12):1165

Hannibal J, Hindersson P, Knudsen SM, Georg B, Fahrenkrug J (2002) The photopigment melanopsin is exclusively present in pituitary adenylate cyclase-activating polypeptide-containing retinal ganglion cells of the retinohypothalamic tract. J Neurosci. 22(1):RC191

Hattar S, Liao HW, Takao M, Berson DM, Yau KW (2002) Melanopsin-containing retinal ganglion cells: architecture, projections, and intrinsic photosensitivity. Science 295 (5557):1065–1070

Haves P, Linden PF, Da Graca G Carrilho (2004) Use of simulation in the design of a large, naturally ventilated office building. Building Serv Eng Res Technol 2004(25):211

Heerwagaen JH, Orians GH (1986) Adaptations to windowlessness: A study of the use of visual décor in windowed and windowless offices. Envir Behav 18(5):623–639

Heschong L (2012) Heschong Mahone Group. Daylight metrics. California Energy Commission. Publication number: CEC-500-2012-053

Hopkinson RG, Bradley RC (1960) Glare from very large sources. Illuminating Eng 55:288–297

Hoyt T, Schiavon S, Piccioli A, Moon D, Steinfeld K (2014) CBE thermal comfort tool.Berkeley: Center for the Built Environment, University of California 2012e2014. http://smap.cbe. berkeley.edu/comforttool/ last Accessed 20 June 14

Huang J, Franconi E (1999) Commercial heating and cooling loads component loadanalysis. Building technologies department, Lawrence Berkeley NationalLaboratory; 1999. LBNL-37208

IEA SHC Task 21 (2000) Daylight in buildings. A source book on daylightingsystems and components. A report of IEA SHC Task 21/ ECBCS Annex 29, July 2000. https://buildings.lbl. gov/sites/all/files/daylight-in-buildings.pdf

IESNA (2016) Lighting handbook. In: DiLaura D, Houser K, Mistrick R, Steffy G (eds) ISBN # 978-0-87995-241-9

Illuminating Engineering Society of North America. Approved Method: IES Spatial Daylight Autonomy (sDA) and Annual Sunlight Exposure (ASE) (2012) IES LM-83-12

Inanici M, Brennan M, Clark E (2015). Multi-spectral Lighting Simulations: computing circadian light. International Building Performance Simulation Association (IBPSA) 2015 Conference, Hyderabad, India, December 7–9, 2015

International Well Building Institute (2016) WELL Building Standard (v1). May 2016

ISO 7730 (2005). Ergonomics of the thermal environment analytical determinationand interpretation of thermal comfort using calculation of the PMV and PPDindices and local thermal comfort criteria. 3rd version. Geneva: InternationalOrganization for Standardization; 2005

Jakubiec JA, Reinhart CF (2011) DIVA 2.0: Integrating Daylighting And Thermal Simulations Using Rhinoceros 3D, DAYSIM and EnergyPlus. In: Proceedings of building simulation 2011:12th conference of international building performance simulation association, Sydney, 14–16 November

Jakubiec JA, Reinhart CF (2012) The 'adaptive zone'—A concept for assessing discomfort glare throughout daylit spaces." Lighting ResTechnol 44(2): 149–70, Originally published online October, 2011

Jakubiec JA, Reinhart CF, Van Den Wymelenberg K (2016) Towards an integrated framework for predicting visual comfort conditions from luminance-based metrics in perimeter daylit spaces. In: Proceedings of bs2015: 14th conference of international building performance simulation association, Hyderabad, India, Dec. 7–9, 2015

Kaplan R (1993) The role of nature in the context of the workplace. Landscape Urban Planning 26:193–201

Kellert Heerwagen (2008). Biophilic design: the theory, science and practice of bringing buildings to life. p 432. Wiley. ISBN: 978-0-470-16334-4

Khalsa SBS, Jewett ME, Cajochen C, Czeisler CA (2003) A phase response curve to single bright light pulses in human subjects. J Physiol 15:945–952

Klepeis NE, Nelson WC, Ott WR, Robinson J, Tsang AM, Switzer P, Behar JV, Hern S, Engelmann W (2001) The national human activity pattern survey (NHAPS): a resource for assessing exposure to environmental pollutants. J Expos Analysis Environ Epidem 11(3): 231–252

Konis K (2012) A method for measurement of transient discomfort glare conditions and occupant shade control behavior in the field using low-cost CCD cameras. American solar energy society (ASES) National solar conference, Denver, Colorado, May

Konis K (2016) A novel circadian daylight metric for building design and evaluation. building and environment, Vol 113, 15 February 2017, Pages 22–38 http://www.sciencedirect.com/science/article/pii/S0360132316304498 http://dx.doi.org/10.1016/j.buildenv.2016.11.025

Lam WMC (1977) Perception and lighting as formgivers for architecture. McGraw-Hill, Inc Lark Spectral Lighting. http://faculty.washington.edu/inanici/Lark/Lark_home_page.html Accessed 25 July 2016

Lighting Research Center (2016) Circadian stimulus calculator. http://www.lrc.rpi.edu/programs/lightHealth/ Accessed 8 June 2016

Lockley SW, Brainard GC, Czeisler CA. (2003) High sensitivity of the human circadian melatonin rhythm to resetting by short wavelength light J Clin Endocrinol Metab 88(9):4502–4505

Lucas RJ, Peirson SN, Berson DM, Brown TM, Cooper HM et al (2014) Measuring and using light in the melanopsin age. Trends Neurosci 31(1):1–9

Lucas et al (2016) Excel-based melanopic illuminance calculator. http://lucasgroup.lab.ls.manchester.ac.uk/research/measuringmelanopicilluminance/ Accessed 25 July 2016

Lynes JA (1977) Discomfort glare and visual distraction. Light Res Technol 9:51–52

Mackey C (2015) Pan climatic humans: shaping thermal habits in an unconditioned society. Thesis: M. Arch., Massachusetts Institute of Technology, Department of Architecture, 2015. http://hdl.handle.net/1721.1/99261

Mardaljevic J (2006) CIBSE national conference 2006: engineering the future. 21–22 March 2006, Oval Cricket Ground, London, UK

McConahey E, Haves P, Chirst T (2002) The integration of engineering and architecture: a perspective on natural ventilation for the new san francisco federal building In: 2002 ACEEE summer study on energy efficiency in buildings. Asilomar, California, USA, 2002. https://eetd.lbl.gov/sites/all/files/publications/lbnl-51134.pdf

Moon P, Spencer DE (1942) Illumination from a non-uniform sky. Illum, Eng

Nabil A, Mardaljevic J (2005) Useful daylight illuminance: a new paradigm to access daylight in buildings. Ligh Res Technol 37(1):41–59

Ne'eman E, Hopkinson RG (1970) Critical minimum acceptable window size: A study of window design and provision of a view. Light Res Technol 2:17–27

Peña R (2014) Living proof. The Bullitt Center, High performance building case study. http://neea.org/docs/default-source/default-document-library/living-proof—bullitt-center-case-study.pdf?sfvrsn=6

Phipps-Nelson J, Redman JR, Dijk D-JJ, Rajaratnam SM (2003) Daytime exposure to bright light, as compared to dim light, decreases sleepiness and improves psychomotor vigilance performance. Sleep 26(6):695–700

Provencio I, Rodriguez IR, Jiang G, Hayes WP, Moreira EF, Rollag MD (2000) A novel human opsin in the inner retina. J Neurosci 20(2):600–605

Rea and Bierman (2016) A new rationale for setting light source luminous efficacy requirements. Light Res Technol September 10, 2016 doi:10.1177/1477153516668230

Rea MS et al (2012) Modelling the spectral sensitivity of the human circadian system. Ligh Res Technol 44(4):386–396

Reinhart CF (2002) Lightswitch 2002: a model for manual control of electric lighting and blinds. Solar Ener 77(1): 15–28 doi:10.1016/j.solener.2004.04.003

Reinhart C (2015) Opinion: Climate-based daylighting metrics in LEEDv4-A fragile progress. Light Res Technol 47(4):388

Reinhart CF, Herkel S (2000) The simulation of annual daylight illuminance distributions-a state of the art comparison of six RADIANCE based methods. Energy Build 32(2):167–187. doi:10.1016/S0378-7788(00)00042-6

Rogers Z (2006) Daylighting metric development using daylight autonomy calculations in the sensor placement optimization tool. Boulder, Colorado, USA: Architectural Energy Corporation. https://www.daylightinginnovations.com/system/public_assets/original/SPOT_Daylight%20Autonomy%20Report.pdf

Ruuger M, Gordijn MCM, Beersma DGM, de Vries B, Daan S (2006) Time-of-day-dependent effects of bright light exposure on human psychophysiology: comparison of daytime and nighttime exposure. Am J Physiol-Reg I 290(5)

Sadeghipour R, Mostapha PM (2013) Ladybug: a parametric environmental plugin for grasshopper to help designers create an environmentally-conscious design. In: Proceedings of the 13th International IBPSA Conference Held in Lyon, France Aug 25–30th. http://www.ibpsa.org/proceedings/BS2013/p_2499.pdf

Selkowitz SE, Kim J-J, Navvab M, Winkelmann FC (1982) The DOE-2 and superlite daylighting programs. LBNL report number LBL-14569. https://eetd.lbl.gov/node/50815

Shehabi A, DeForest N, McNeil A, Masanet E, Lee ES, Milliron D (2013) US energy savings potential from dynamic daylighting control glazings. Energy Build 66(2013):415–423

Suk JY, Schiler M, Kensek K (2013) Development of new daylight glare analysis methodology using absolute glare factor and relative glare factor. Energy Buildings 64:113–122

Suk Jae, Schiler M, Kensek K (2016) Absolute glare factor and relative glare factor based metric: predicting and quantifying levels of daylight glare in office space. Energy Build 130(15):8–19

Thapan K, Arendt J, Skene DJ (2001) An action spectrum for melatonin suppression: evidence for a novel non-rod, non-cone photoreceptor system in humans. J Physiol 535:261–267

Torcellini P et al (2006) Lessons learned from field evaluation of six high-performance buildings. NREL/CP-550-36290

Tregenza PR, Waters IM (1983) Daylight coefficients. Light Res Technol 15(2):65–71. doi:10.1177/096032718301500201

Tuaycharoen N, Tregenza PR (2007) View and discomfort glare from windows. Light Res Technol 39(2):185–200

Ulrich RS, Simons RF, Losito BD, Fiorito E, Miles MA, Zelson M (1991) Stress recovery during exposure to natural and urban environments. J Environ Psychol 11(3):201–230

U.S. Green Building Council (2009). LEED 2009 for new construction and major renovations. Reference guide

U.S. Department of Energy (DOE) (2015). A common definition for zero energy buildings. http://energy.gov/sites/prod/files/2015/09/f26/bto_common_definition_zero_energy_buildings_093015.pdf

U.S. Green Building Council. (2016). LEED v4 for building design and construction. April 5, 2016. 161 pages. http://in.usgbc.org/sites/default/files/LEED%20v4%20BDC_04.05.16_current.pdf

Van Den Wymelenberg KG (2012) Evaluating human visual preference and performance in an office environment using luminance-based metrics. PhD Dissertation, University of Washington. ProQuest, LLC

Van Den Wymelenberg K, Inanici MN (2014) A critical investigation of common lighting design metrics for predicting human visual comfort in offices with daylight. LEUKOS 10(3):145–164

Veitch JA, Farley KMJ (2001) A room with a view: a review of the effects of windows on work and well-being. National Research Council of Canada (NRCC) technical report IRC-RR-136. http://nparc.cisti-icist.nrccnrc.gc.ca/eng/view/object/?id=ca18fccf-3ac9-4190-92d9-dc6cbbca7a98

Vos JJ (1984) Disability glare—a state of the art report. CIE J. 3:39–53

Ward G (1992) RADIANCE visual comfort calculation.http://radsite.lbl.gov/radiance/refer/Notes/glare.html

Ward GL, Shakespeare R (1998) Rendering with radiance. Morgan Kaufmann, 1998. ISBN 1-55860-499-5

Weinold J (2009) Dynamic daylight glare evaluation. building simulation. Eleventh International IBPSA Conference. Glasgow Scotland, July 27–30, 2009. http://www.ibpsa.org/proceedings/BS2009/BS09_0944_951.pdf

Wienold J, Christoffersen J (2006) Evaluation methods and development of a new glare prediction model for daylight environments with the use of CCD cameras. Energy Build 38(7):743–757

Wilson EO (1993) Biophilia. Harvard University Press, ISBN 9780674074422

Zelinski EL, Deibel SH, McDonald RJ (2014) The trouble with circadian clock dysfunction: multiple deleterious effects on the brain and body. Neurosci Biobehavioral Rev 40:80–101

Chapter 3
Innovative Daylighting Systems

3.1 Introduction

The building facade and perimeter zone represents a complex design integration challenge due to the diverse array of design and functional requirements paired with the increasing number of energy and environmental objectives set by design teams seeking to achieve a low-energy design concept that simultaneously supports a high level of indoor environmental quality. As designers seek to integrate daylighting within an efficient whole-building energy strategy, it is challenging to manage trade-offs between performance objectives such as envelope thermal performance, lighting and HVAC energy demand with human factors such as visual comfort, daylight availability, visual connection to the outdoors, and personal control. This requires an integrated approach to the application of technology, informed at a fundamental level by empirical knowledge of end-user needs. The following sections define four key areas of advancement for daylighting technologies that have the potential to help enable high-performance daylit buildings with enhanced indoor environmental quality.

1. From Simple to Complex Fenestration Systems
2. From Static to Dynamic Systems
3. From Integrated to Interconnected Systems
4. From Closed-loop to Human-in-the-loop Systems

These four dimensions of evolving daylighting design solutions seem to suggest a trend from simple passive solutions to complex active solutions with potential negative connotations. We subscribe to the philosophy that "everything should be as simple as possible, but no simpler." Some daylighting solutions may need to be complex because people and tasks have ever changing needs, and because changing climatic conditions require dynamic responses. But we believe the apparent complexity being added is not unlike many other aspects of daily life, where the advent of a variety of hardware and software solutions and services are already rapidly

© Springer International Publishing Switzerland 2017
K. Konis and S. Selkowitz, *Effective Daylighting with High-Performance Facades*, Green Energy and Technology, DOI 10.1007/978-3-319-39463-3_3

changing living and working environments. In the case of daylighting solutions for buildings we have the opportunity to try to create design process and technology solutions that are deployable, scalable, robust, and economic.

Case study and theoretical examples are used to discuss promising glazing and facade technologies supported by emerging digital infrastructure, sensing and controls that are broadly applicable for new and existing buildings. In each section, examples illustrate that occupant experience and requirements serve as a key driver for technology and design solutions development.

3.2 From Simple to Complex Fenestration Systems

Glazing systems in modern, energy-efficient buildings consist of a number of individual components (glass layers, gas layers, frames, spacers, and dividers). Changes in building codes over the past 40 years track slow but steady progress in thermally improved glazing technology and a variety of supporting optical technologies for improved control of solar heat gains. In regard to thermal performance, the AEC industry has moved from a standard of single glazing with a $U \sim 6$ W/m^{2}K to triple glazings with gas fills and two low-E coatings with $U < 0.6$ W/m^{2}K. Market impact is slowed by cost and in some cases by the need to redesign window sash, frames and hardware to accommodate heavier, larger Insulating Glass Units (IGUs). Emerging evacuated glazings and other novel thin multilayer glazing and coating design options offer even better performance at lower weight and cost (Jelle et al. 2012a, b). And in regard to solar control, spectrally selective glazings (Fig. 3.1), with multilayer thin film coatings, now offer very nearly ideal control of the visible and near IR spectrum- transmitting over 90% of the visible light and less than 10% of the near IR, which is ideal for cooling dominated climates where daylight admittance is important.

In addition to thermal performance and solar control, objectives for fenestration design are increasingly driven by visual comfort, view, daylight sufficiency and the quality of interior daylight distribution. Efforts to address these additional objectives has led to the development of a broad range of systems that actively and passively manipulate incident light in ways that are more complex than with conventional glazing. The defining characteristic of most glazings (clear, tinted, reflective, low-E) is that they are specular, they do not change the angular direction of the light passing through them. All other facade systems that can alter transmitted light direction are termed Complex Fenestration Systems (CFS). Such systems include optical light redirecting systems, mirrored louver systems, frits, prismatic window films, diffusing glazings, shade fabrics, and angular selective screens. In contrast to optical data for specular glass layers, where transmission and reflection can be defined through simple mathematical functions, complex fenestration systems typically involve assemblies of multiple non-specular layers, each with its own unique transmittance, absorptance and reflectance depending on the angle of incidence of the light source.

Fig. 3.1 Spectral
transmittance *curves* for
glazing with three different
types of low-emittance
coatings. *Image credit*
Lawrence Berkeley National
Laboratory

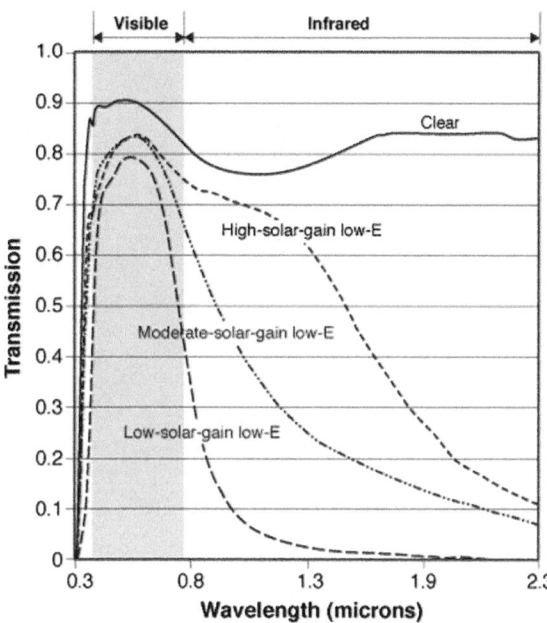

To accurately model the daylighting performance of CFS, light rays must be propagated and traced as they pass through each layer and their path altered depending upon the nature of each layer. The most powerful method to carry out this approach is with ray-tracing software, for example the methods that have been developed for the lighting simulation software Radiance (Ward et al. 2011), which utilizes a backwards ray tracer and Bidirectional Scattering Distribution Function (BSDF) data for each material layer. As shown in Fig. 3.2, a BSDF is a set of hemispherical luminous coefficients defined by paired incident and outgoing angles (Ward et al. 2011). In the term BSDF, "scattering" refers to both transmittance and reflectance. For each incoming ray arriving at any arbitrary incidence angle, the "distribution" is used to describe the pattern of the outgoing transmitted rays over every outgoing angle in the full hemisphere.

By convention most BSDF data is measured with 145 incoming light directions and 145 outgoing light directions to create a file with $145 \times 145 = 21{,}025$ entries and stored in a standardized .xml file format that can be viewed interactively using the BSDF viewer utility (Fig. 3.3). The viewer utility allows the user to load a BSDF xml file and view the outgoing distribution for user-selectable incident directions and look at transmission or reflection for front or back. An example BSDF visualization for the light redirecting louver system (Figs. 3.4 and 3.5) is provided in Fig. 3.3. Figure 3.3 shows the resulting distribution of transmitted visible light (right hemisphere) for one incident sky patch highlighted in yellow (left hemisphere).

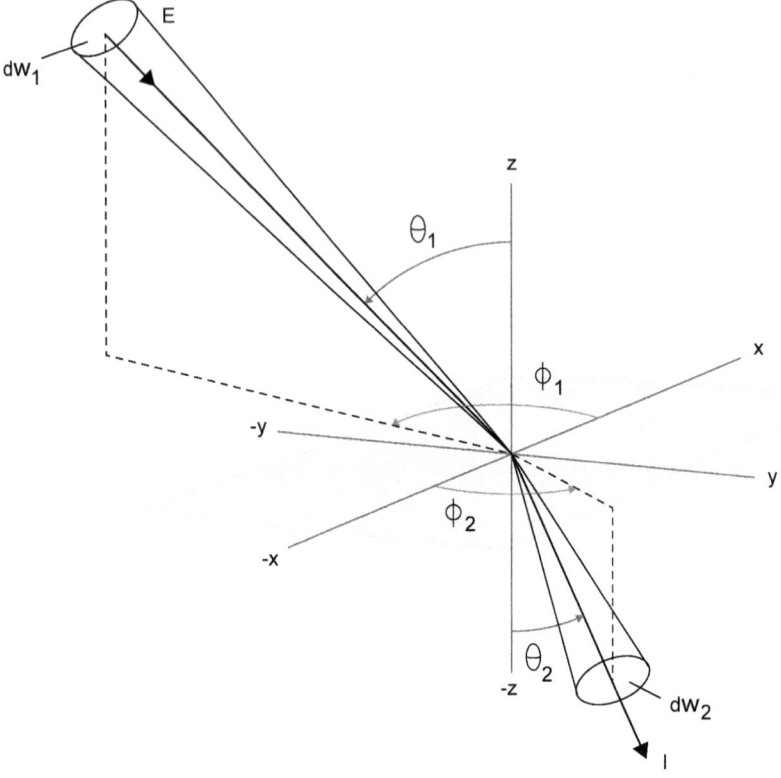

Fig. 3.2 Coordinate system for bidirectional measurements, showing sample input and output rays in the transmission mode. Image drawn by Sue Long Lee (after Ward et al. 2011)

Traditional optical data consisted of a single transmittance value measured at normal incidence to the glass. The growing availability of BSDF data with $\sim 21,000$ data entries for a given glazing/shading system allows designers to more accurately simulate and compare the performance of various CFS options during design and to evaluate and refine the unique daylighting behavior of each alternate CFS design. BSDF data is generated by characterizing a given material (i.e. obtaining bi-directional optical measurements) using an optical measuring system such as a scanning goniophotometer (SOURCE) or is developed with computational means through software simulation using the Radiance program *genBSDF* (reference Ward) if the geometry and materials are well described. The software tool Berkeley Lab WINDOW includes a BSDF library of existing daylighting technologies and provides the capability for users to describe any arbitrary assembly of glazing, shading, and other optically-complex coplanar layers which is then processed to generate a unique BSDF characterization of the assembly (see Chap. 4). The BSDF generated for a given system can then be used

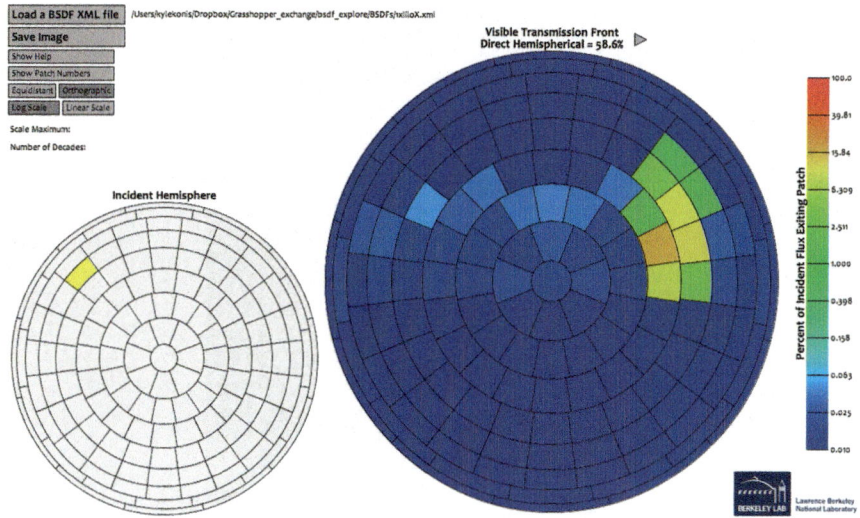

Fig. 3.3 Example visualization of BSDF data for a light redirecting shading system showing output distribution (*right*) for a single input direction (*left*). *Image credit* LBNL

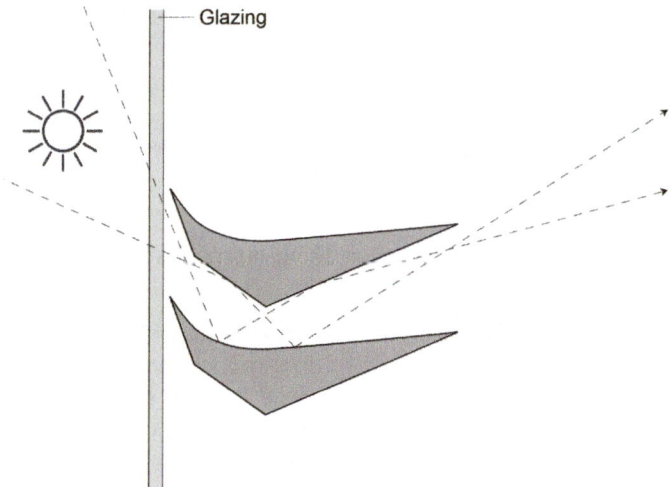

Fig. 3.4 The patented LightLouver™ reflective slat design redirects all sunlight above a 5° altitude angle upward onto the ceiling of the daylit space, providing ambient lighting and eliminating direct sunlight on work surfaces (http://lightlouver.com/lightlouver-description/). Image drawn by Sue Long Lee

as a material description in Radiance to perform accurate, computationally efficient daylighting simulations to determine illuminance levels or assess the potential for glare. A simulation example is provided in the following section (Fig. 3.6).

Fig. 3.5 Close up profile view of a LightLouver™ daylighting system unit during installation. After installation the unit is parallel to the window glazing (inboard side). The LightLouver™ daylighting system units redirect sunlight deeper into the office space to reflect off ceilings and provide ambient lighting. *Image credit* Dennis Schroeder/NREL

3.2.1   Optical Light Redirecting Systems (OLS)

A square meter (m^2) of direct beam sunlight contains about 100,000 lumens. If these could be evenly distributed without loss over the interior space behind the window they could provide 500 lux over 200 m^2 of floor space. Even with optical losses the window could potentially light the room 12–15 m from the window. But direct sunlight normally falls to the floor by the window and diffuse daylight is reduced rapidly as one moves away from the window. By utilizing geometrically designed specular surfaces or refractive optics, optical sunlight redirecting systems integrated into the overhead "daylight" zone of the building facade present the potential to enlarge the daylit area of the floor plate by redirecting the sunlight incident on the window deeper into the space than conventional shading systems, which tend to absorb, back reflect or diffuse the sunlight. More significantly, by developing system geometry to redirect daylight primarily to the ceiling within the space, Optical Light redirecting Systems (OLS) have the potential to avoid the glare conditions commonly produced by conventional facade shading systems which diffuse and distribute significant amounts of daylight below eye level into the

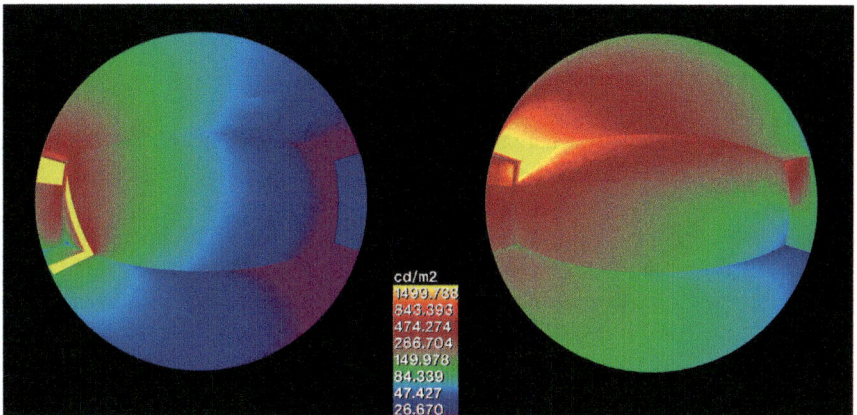

Fig. 3.6 Radiance simulation comparing the luminance distribution of an unshaded upper daylight zone window (*left*) with the LightLouver™ CFS (*right*) using BSDF data to model the daylight redirecting behavior of the LightLouver™ on September 21, 11:00 AM in San Francisco, CA climate with clear sky conditions

occupant's field of view. Since these systems require direct sunlight they are most appropriate on southeast to south to southwest facades and in climates with a high fraction of direct sunlight. We explore the optical functionality of two classes of devices- optically reflective devices and optically refractive devices.

3.2.1.1 Reflective OLS

The louver system discussed above in explaining BSDFs is a good example of an optically reflective device. The device uses a combination of carefully designed optical geometries and high reflectance surfaces and is installed just inside an upper clerestory windows. Figure 3.6 shows a simple example of a transverse section perspectival view of a south-facing facade, comparing the daylight distribution of a conventional glazed upper window aperture (left) with the LightLouver™ OLS (right). A falsecolor mapping is applied to visualize the resulting luminance (cd/m^2) distribution of both conditions. Figure 3.6 shows that by redirecting direct sun towards the ceiling (right), the LightLouver™ effectively eliminates penetration of direct beam to the floor, reduces the luminance contrast of the upper daylighting aperture with surrounding surfaces, and distributes significantly more daylight towards the ceiling and the back of the room. The result is a reduction in glare and occupant solar overheating in the perimeter zone, along with reduced need for electrical lighting away from the perimeter zone.

An additional limitation is performance under non-clear sky conditions (e.g. cloudy and overcast) where performance is significantly reduced relative to conventional (unshaded) windows. A detailed study documenting the measured daylighting potential of the LightLouver™ static optical louver system under real sun and sky conditions is provided by Konis and Lee (2015).

The fenestration strategy for the National Renewable Energy Laboratory (NREL) Research Support Facility (RSF) south-facing facade (Figs. 3.7 and 3.8) illustrates a design strategy where the window is subdivided vertically into an upper daylight zone and lower view zone, where the upper zone incorporates a static OLS (LightLouver™). The singular function of the upper daylight zone is to redirect incident direct beam sunlight towards the ceiling to daylight the interior open-plan work zones deep within the building floor plate without creating glare (Fig. 3.9). For solar control, an exterior overhang and vertical shade fins are positioned under the OLS to shade the view zone but to not block direct beam to the OLS. The view zone is subdivided horizontally to incorporate a manually controlled operable window enabling occupants to have greater personal control over building ventilation and personal comfort. Notably, no interior shading devices are used since the view windows are not large and are externally shaded.

Due to the sunny climate and the daylight redirecting capabilities of the OLS, the window can achieve the daylighting requirements using relatively little facade area (30% WWR) compared with conventional window and shading systems. Consequently, more of the facade can remain opaque, enabling higher thermal

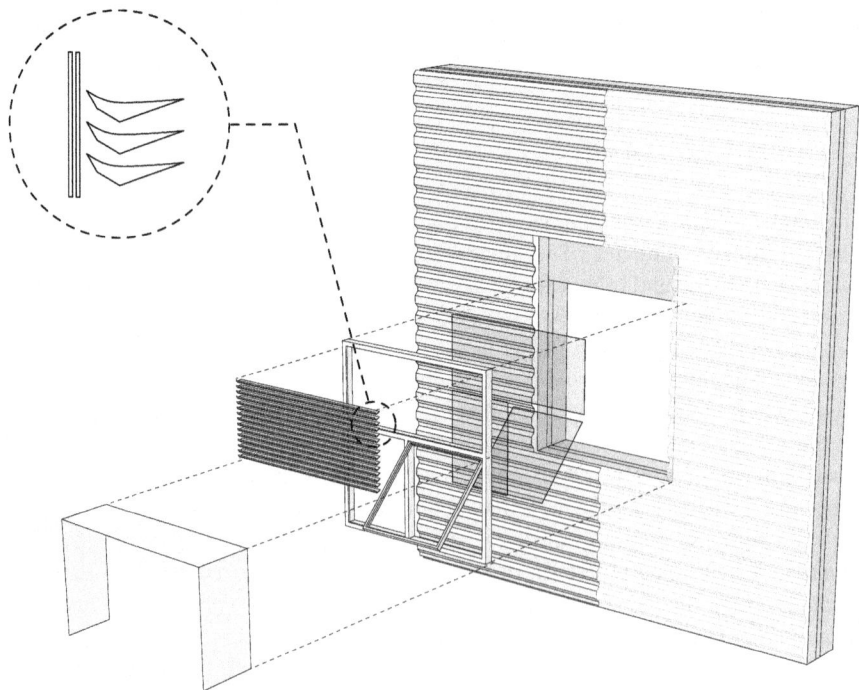

Fig. 3.7 Exploded axonometric of the research support facility (RSF) at the National Renewable Energy Laboratory (NREL). Image drawn by Sue Long Lee

Fig. 3.8 NREL RSF south elevation showing functionally subdivided windows. *Image credit* Pat Corkery/NREL

Fig. 3.9 The LightLouver™ daylighting system units installed in the daylight windows (*right side of image*) redirect sunlight deep into the office space to provide ambient lighting. *Image credit* Dennis Schroeder/NREL

performance and the performance of additional energetic/environmental functions applicable to sections of opaque wall. In this case, much of the opaque area on the RSF south facade incorporates a transpired solar collector to passively pre-heat ventilation air during the heating season. A complete case study description of the RSF project is provided in Chap. 5.

3.2.1.2 Optically Refractive Films and Coatings

In addition to 3-dimensional mirror and light shelf reflective structures, 2-dimensional light redirecting coatings and prismatic films are emerging, enabled by various micro and nano-scale fabrication techniques. These 2-dimensional technologies (Fig. 3.10) have the potential to provide useful daylighting 10 m from the perimeter with upper window zone only, at significantly lower claimed cost compared with 3-dimensional alternatives, leading in turn to greater market adoption and energy impacts. As with the refractive optics the goal is to keep the redirected light above the horizontal level. As can be seen in Fig. 3.10 both prismatic devices do a good job of redirecting the transmitted light to the ceiling, compared to the reflective blind. However, effective illuminance distribution varies

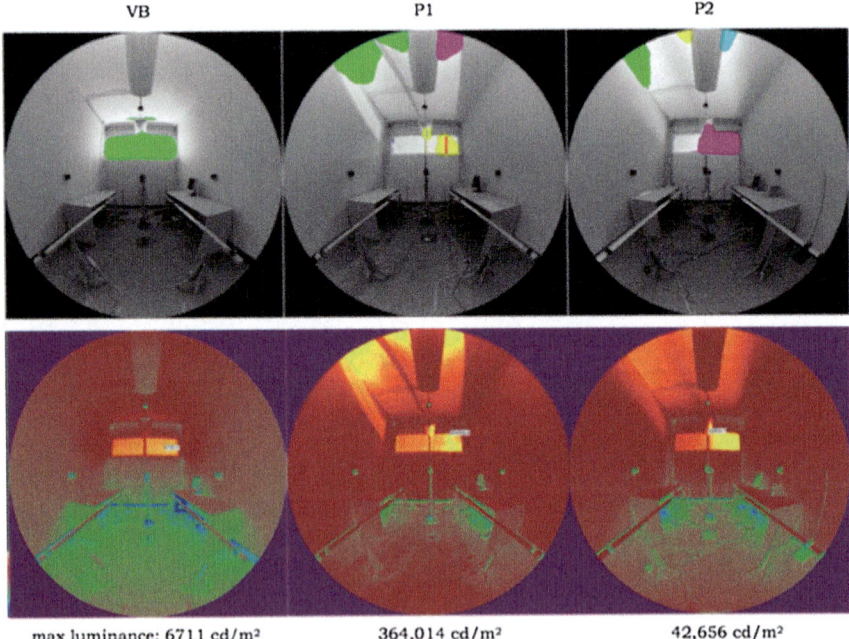

max luminance: 6711 cd/m²   364,014 cd/m²   42,656 cd/m²

Fig. 3.10 Shows three design options. *Left* is reflective venetian blind. *Center* is prismatic v1. *Right* is prismatic v2. *Image Credit* LBNL

as the sun changes altitude and glare control with daily and seasonal changes in sun and sky conditions remains an optimization challenge. Optical designs that let through more light for daylight impact tend to have more glare; designs that manage glare better do not save as much lighting energy.

3.2.2   Angular Selective Glazing Systems

To meet low and ZNE whole-building energy objectives, facade systems are needed that control solar loads and glare while maintaining transmission of useful daylight and views to the outdoors. Angular Selective Glazing Systems (ASGS) present one cost-effective and practical approach to achieving this goal. ASGS are static cellular or louver structures embedded within the facade glazing, where the three-dimensional geometry can be manipulated to block or admit solar radiation for specific sun angles while maintaining transmission of useful daylight and only partially-occluded views to the outdoors (Fig. 3.11). Some degree of angle selectivity can also be achieved with a planar perforated shade where the perforations have angular selective properties.

ASGS offer a range or architectural possibilities to enlarge the daylit zone for new and retrofit applications by enabling larger areas of facade glazing while still meeting solar and glare control requirements. Figure 3.12 presents an example of Microshade™, a micro-scale static louver structure integrated into a double pane Insulating Glazing Unit (IGU). The technology consists of microscopic lamellas, which (for vertical applications) are designed to progressively block increasingly levels of direct sun as solar altitude increases. The shading effect of the microscopic lamellas is similar to that of exterior blinds, except the micro lamellas are too small

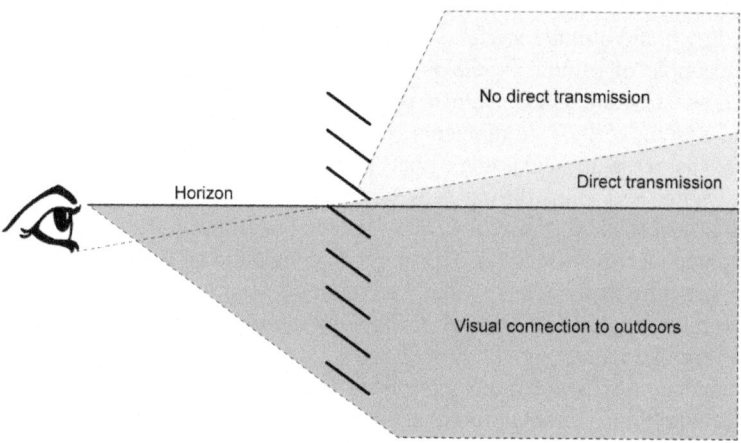

Fig. 3.11 Example cross-sectional view of an angular selective glazing system (ASGS). Image drawn by Sue Long Lee

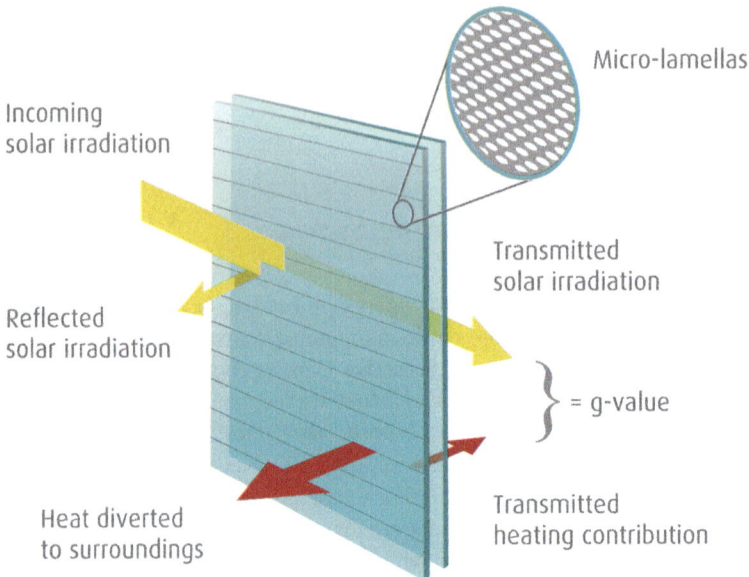

Fig. 3.12 Diagrammatic view of Microshade™, a micro-scale static louver structure integrated into a double pane insulating glazing unit (IGU). *Image credit* MicroShade

for the human eye to see, leading to a more transparent visual effect when viewed from the building interior. The Microshade™ IGU does not distort the color of transmitted daylight compared to a visibly tinted solar control glazing, as shown in Fig. 3.13. As an ASGS, the solar and optical performance of the technology change with the incident angle of solar radiation. Tables 3.1, 3.2 and 3.3 present the g-Value, direct solar transmittance, and direct visible light transmittance respectively for the MicroShade (type = MS-A[1] Vertical, Two-Layer IGU) for a range of solar azimuth and altitude angles.

An example of cellular ASGS is ClearShade™. The technology consists of a honeycomb structure made from a polymer composition encapsulated within a standard system of IGU components (Fig. 3.14). The technology is conceptually designed as a static technology that performs dynamically by progressively reducing solar heat gains during peak hours while maintaining visible light transmission at levels above those of heavily tinted solar control glazings and films, or highly-tinted electrochromic (EC) glazings. In contrast to an EC or typical IGU, the cellular structure of the ClearShade™ technology results in increasingly occluded views to the outdoors as the angle of the view increases from the window surface normal (Fig. 3.15).

In practice, ASGS are rarely optimized for specific facade orientations, local climatic conditions, internal room geometries or specific lighting requirements.

---

[1]Data from http://www.microshade.net/media/914/ms_a-vertical_en_2013.pdf.

Fig. 3.13 Comparison of view to outdoors looking through MicroShade™ (installed in *right window*) versus solar control glass (*door* and *top window*). *Image credit* MicroShade

Table 3.1 g-Value

		Solar altitude (°)					
		0	15	30	45	60	75
Azimuth	0	0.39	0.35	0.29	0.21	0.09	0.03
	15	0.38	0.33	0.28	0.21	0.09	0.03
	30	0.35	0.31	0.26	0.18	0.07	0.03
	45	0.3	0.26	0.21	0.14	0.05	0.02
	60	0.2	0.17	0.13	0.07	0.03	0.02
	75	0.03	0.03	0.03	0.02	0.02	0.01

Table 3.2 Direct solar transmittance

		Solar altitude (°)					
		0	15	30	45	60	75
Azimuth	0	0.33	0.29	0.24	0.17	0.05	0
	15	0.32	0.28	0.23	0.16	0.05	0
	30	0.29	0.26	0.21	0.14	0.04	0
	45	0.24	0.21	0.17	0.1	0.02	0
	60	0.15	0.13	0.09	0.04	0	0
	75	0	0	0	0	0	0

Table 3.3 Visible light transmittance

		Solar altitude (°)					
		0	15	30	45	60	75
Azimuth	0	0.48	0.41	0.34	0.24	0.08	0
	15	0.46	0.4	0.33	0.23	0.07	0
	30	0.42	0.37	0.29	0.19	0.05	0
	45	0.35	0.3	0.24	0.14	0.02	0
	60	0.22	0.18	0.13	0.05	0	0
	75	0.01	0	0	0	0	0

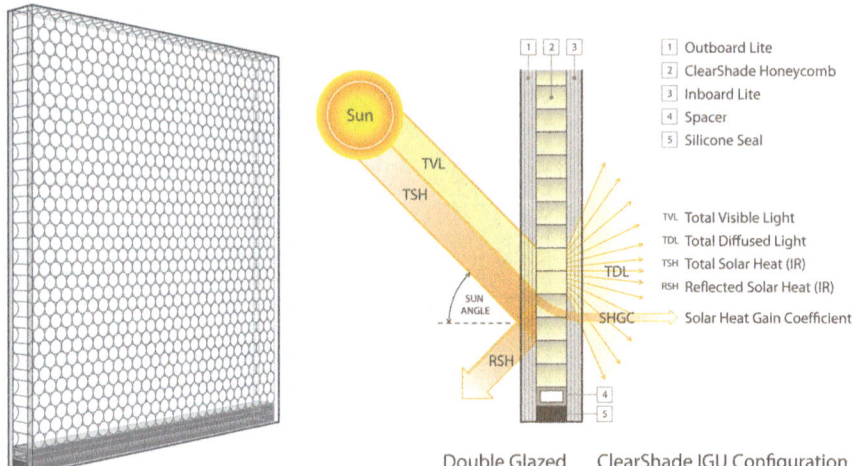

Fig. 3.14 Double glazed ClearShade IGU configuration showing perspective view (*left*) and section view (*right*). *Image Credit* Panelite

Developments in the application of BSDF data, simulation-based "form-finding," and mass customization are helping to enable the development and refinement of complex fenestration systems that can be "tuned" to context-specific solar and climatic conditions in and unique performance requirements.

Figure 3.16 shows early prototype geometry for an ASGS developed for urban Los Angeles (CA climate zone 9) for a SW-facing facade. The system geometry consists of hexagonally-packed cylinders where the extrusion length and direction (relative to the facade normal) is refined through simulation using objective functions for (1) minimizing solar loads during the cooling season, (2) glare control, (3) admission of useful daylight, and (4) minimizing occlusion of view to the outdoors. Rapid prototyping of the system using a consumer-level 3D printer (Figs. 3.17 and 3.18) enables prospective geometries to be physically examined to evaluate characteristics such as directionally-dependent visual occlusion, solar and glare control (Fig. 3.19). As of now 3-D is probably not scalable to produce large

Fig. 3.15 Example views available looking through ClearShade™ installed in exterior curtain wall glazing at INV Management, NJ. Gluckman Mayner architects. *Image Credit* Panelite

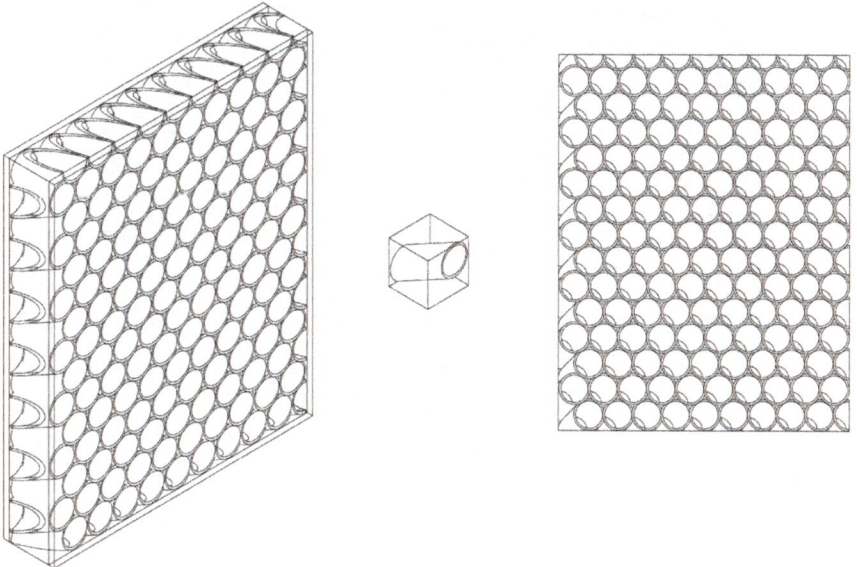

Fig. 3.16 Schematic view of hexagonally-packed diagonally-extruded cylinder geometry

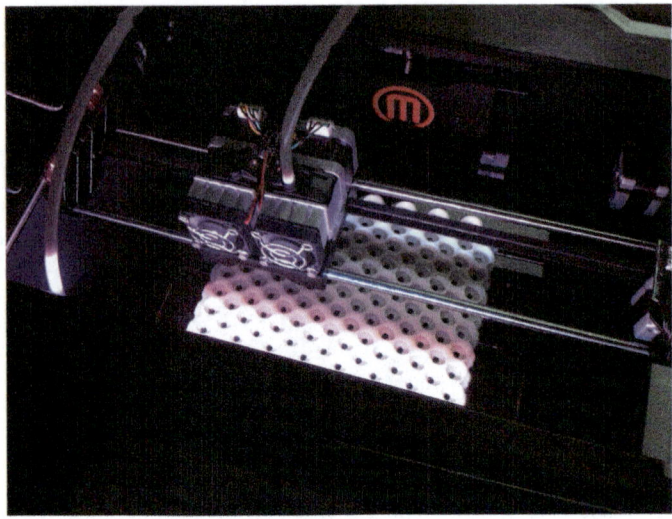

Fig. 3.17 3D-printing of the ASGS screen structure using consumer Makerbot 3D printer

Fig. 3.18 View of the completed "print" of the ASGS screen structure

quantities of this type of product but it is well suited for prototype development and optimization, and this approach has also been used to make molds for mass production. The ability to customize shading solutions with tradeoffs between solar control and daylight admittance that are customized for orientations, latitude and climate would be a powerful capability in the hands of architects and engineers.

Fig. 3.19 Inspection of the ASGS screen structure sample under real sun and sky conditions

Fig. 3.20 Photograph of light-diffusing capillary slab integrated within standard double-glazing IGU (Okalux). The capillary slab can be incorporated into a range of standard IGU configurations (e.g. double and triple glazing, gas fill, with low-e coatings). *Image credit* Okalux

At a smaller scale, capillary systems, such as the light-diffusing capillary slab technology integrated in various Okalux technologies (Okalux 2016), offer the potential to more evenly distribute transmitted daylight and control visual contrast relative to conventional window glazing systems. The technology consists of

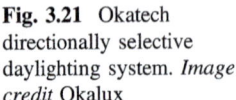

Fig. 3.21 Okatech
directionally selective
daylighting system. *Image
credit* Okalux

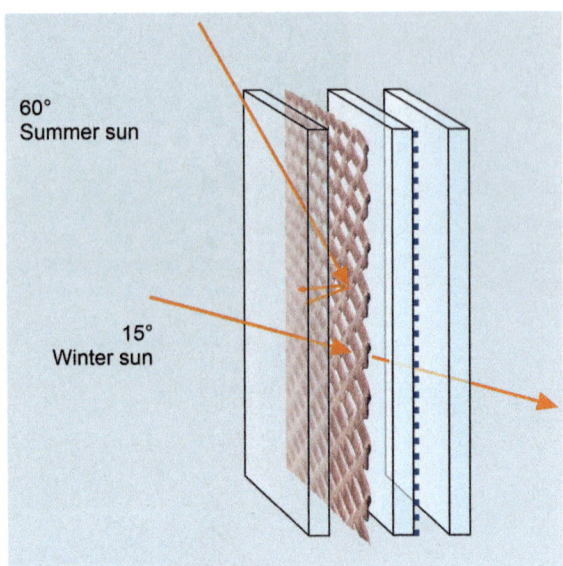

translucent capillary tubes (Fig. 3.20) which function both to intercept and diffuse
incident beam sunlight as well as to prevent the convection of air within the cavity
to reduce heat transmission. The capillary slab can be incorporated into a range of
standard IGU configurations (e.g. double and triple glazing, gas fill, with low-e
coatings).

Finally, interlayers of expanded metal, or wire mesh can be integrated into IGUs
to impart directional selectivity. Figure 3.21 shows a schematic view of Okatech
insulating glass with expanded metal. Directional selectivity can be "tuned" to
specific facade requirements by varying the 3-dimensional geometry of the
expanded metal. As shown in Fig. 3.21, the metal interlayer functions to block
increasing levels of direct beam sunlight as the angle of incidence increases.

3.2.3 Ceramic Frits

Ceramic frits (Fig. 3.22) represent another category of CFS that enable static
adjustment of solar heat, daylight transmission, privacy and visual connection to the
outdoors. Ceramic frit typically consists of finely ground glass mixed with inorganic
pigments to produce a desired color. Frit is typically applied to the surface of the
glass using a silk-screening process and then heated within a tempering furnace to
create a durable permanent coating. Newer processes use the equivalent of "ink-jet

Fig. 3.22 Example ceramic frit pattern on glazing sample

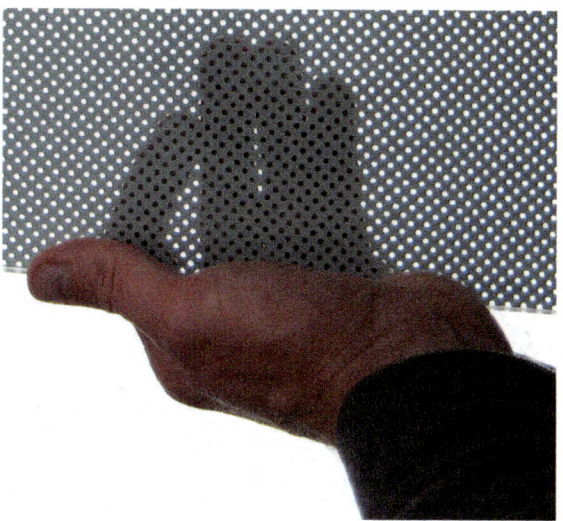

printing" to produce photographic images. Geometric patterns can be fine-grained or of larger dimensions, e.g. alternative 10 cm wide bands. Depending upon the materials used and the thickness of the coating it can be translucent or opaque. Opaque areas of the pattern act as screens, blocking solar radiation. Fritted glass can also be combined with high-performance coatings and other glass substrates as part of an IGU. From an architectural design perspective, ceramic frits can have a significant effect on the outward appearance of a building, due to the ability to achieve custom patterns and coloration of the exterior glass layer. In practice, designers can create the desired appearance of a fully-glazed "all-glass" facade exterior, while providing more freedom to size window apertures for daylighting, views, and ventilation based on program and environmental performance requirements. Figure 3.23 shows an exterior view of the ceramic frit pattern on the library facade of the John and Frances Angelos Law Center, located in Baltimore, MD. Dissipation of the frit pattern vertically (Fig. 3.24) is intended to achieve solar control requirements while maintaining high levels of daylight transmission through the upper section of the window as well as clear downward views from the lower windows (Fig. 3.25).

While ceramic frits have been commercially available for decades, the emergence of visual scripting tools (e.g. Grasshopper Image Sampler Fig. 3.26) and customized silkscreening techniques and digital printing have enabled design teams to manipulate light transmission with unique patterns (Fig. 3.27). For example, an image of the shadow pattern produced by tree canopy (Fig. 3.28) can be sampled

Fig. 3.23 Ceramic frit pattern on the library facade of the John and Frances Angelos Law Center, located in Baltimore, MD. Over the library facade the frit covers approximately seventy percent of the wall, protecting the interior from solar gain. One-half of the panels are fully fritted, and the other half are coated with a custom gradient frit pattern that alternates a half-floor height every other panel, creating a three-dimensional 'woven' effect. *Image credit* Behnisch Architekten

(Fig. 3.29) and applied as a frit pattern to produce indoor lighting conditions that may more closely mimic the light/dark patterns of the natural environment.

The examples in the above sections illustrate that there are many options to provide different degrees of control of light, view, glare and solar gain as a function of climate and site parameters, other aspects of the glazing and shading system, and occupant requirements for task performance or comfort. The requirement to optimize all of these designs using a wide range of different technologies requires as a starting point that the modeling tools are available to assess performance and that the product properties, e.g. BSDF, are measured for the systems under

Fig. 3.24 Frit pattern for alternating upper and lower window units. *White* indicates ceramic frit, *black* indicates view through glass. *Image credit* Behnisch Architekten

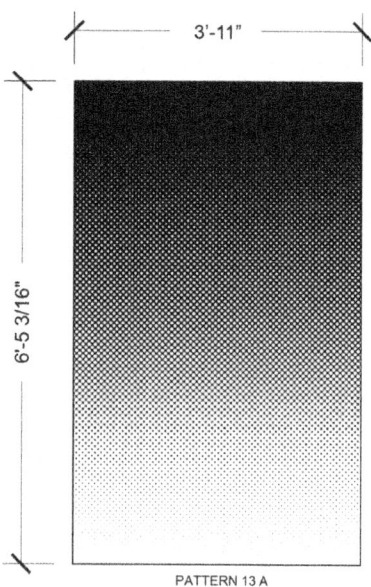

3'-11"

6'-5 3/16"

PATTERN 13 A

Fig. 3.25 Interior view of the library facade of the John and Frances Angelos Law Center showing alternating window pattern, where the application of frit preserves both views downward near desks and clear views to the sky from *upper window* regions. *Image credit* Behnisch Architekten

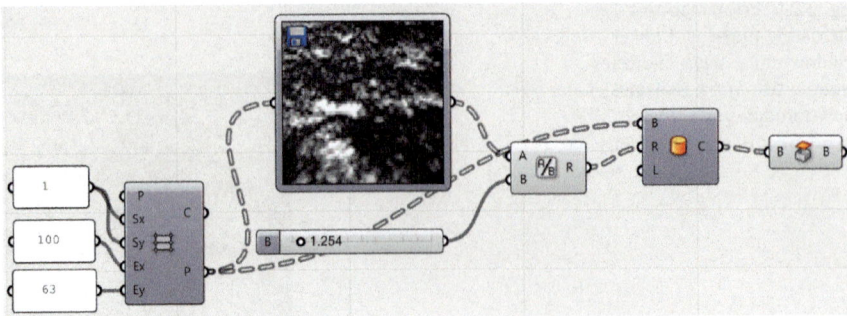

Fig. 3.26 The Grasshopper image sampler component. The image sampler uses the color brightness channel of the image specified by the user to individually size the opaque (fritted) area of each region of the window unit

Fig. 3.27 Resulting frit pattern generated from the Grasshopper image sampler component. *White* indicates ceramic frit, *black* indicates view through glass

consideration. The primary thermal, solar and daylight simulation models are now available but the optical data is only now beginning to be created. The Complex Glazing Data Base currently has only about 400 data entries for the CFS systems whereas the glazing data base holds over 5000 entries representing the majority of all commercially available glazing systems.

Fig. 3.28 Example image of shading pattern created by sidewalk tree canopy in urban Los Angeles

Fig. 3.29 Low-resolution image of shadow pattern used as data input to image sampler component

3.2.4   Building Integrated Photovoltaics (BIPV)

Most ZNE building designs rely on building integrated photovoltaics (BIPV) as a decentralized source of renewable energy to offset site energy consumption over an annual period. For low-rise buildings, the available roof area is often sufficient to meet ZNE if the project is already designed to minimize energy demand. For example, three of the case study projects in this book (Bullitt Center, NREL RSF, and NewActon Nishi) achieve ZNE through the application of rooftop PV. However, the increasing scale of ZNE projects, combined with competing uses of rooftop area (e.g. public space, vegetative roof coverings, mechanical equipment and access ways etc.) leads to the more technically challenging task of integrating PV into the vertical facade of the building. Multiple manufacturers now offer facade systems with PV integrated into opaque curtain wall and a growing number are exploring PV integrated into the view glazing (Jelle et al. 2012b). Technologies include crystalline cells, expanded cells, and amorphous cells. While there is some reduction in output due to the vertical rather than horizontal surface orientation, many of the balance of systems costs can be offset by the curtain wall framing and other elements serving dual roles. The addition of PV to the view glazing or spandrel places additional demand on the design team to fully integrate structure, wiring, power, view etc. in the curtain wall solution, as well as to examine and consider factors of surface orientation optimization and urban overshadowing.

A central facade design challenge is whether to integrate the solar generation panels as spandrel panels (Fig. 3.30), as shading elements above the glazing (Fig. 3.31) or in the glazing itself (Fig. 3.32). Each of these approaches has its pros and cons. There is growing interest in the integration of the PV elements in the vision area of the facade, either as an amorphous "see-through" layer or using thin opaque strips that are embedded in laminated glass covering 20–70% of the overall glass area. In this latter case the appearance from inside is similar to looking through horizontal louvers (Fig. 3.32).

Traditional PV systems convert the more energetic visible rays of sunlight to power so the more power generated the less light is transmitted. The latest generation of glazing integrated PV uses semi transparent layers that generate power primarily by absorbing in the near IR, thus they can still have a reasonably high daylight transmittance. Current products have some color but further research is targeting neutral colors. These are finding early applications to charge flat panel electronics but should eventually find more applications in windows and skylights.

A final note of caution in the interest to use glazed areas to generate electric power for daylighting versus using the glazing to admit daylight is that direct use of daylight, even with the optical losses within windows or skylights and room

Fig. 3.30 PV system integrated into facade glazing and spandrel panel. *Image credit* LBNL

Fig. 3.31 PV system integrated into exterior horizontal shading. *Image credit* LBNL

Fig. 3.32 Expanded photovoltaics cells integrated into glazing. In this type of application, the solar cells provide shading and generate electricity. *Image credit* LBNL

cavities, is still a more efficient pathway to deliver lumens to a space for lighting than via PV to electric light conversion. This is illustrated conceptually in the diagram below (Fig. 3.33), which shows respective system "efficiencies" of 2, 25 and 8% for PV, skylights, and vertical windows. Further improvements in PV efficiency and LED efficacy will shift the numbers but the underlying comparison performance issues should be considered.

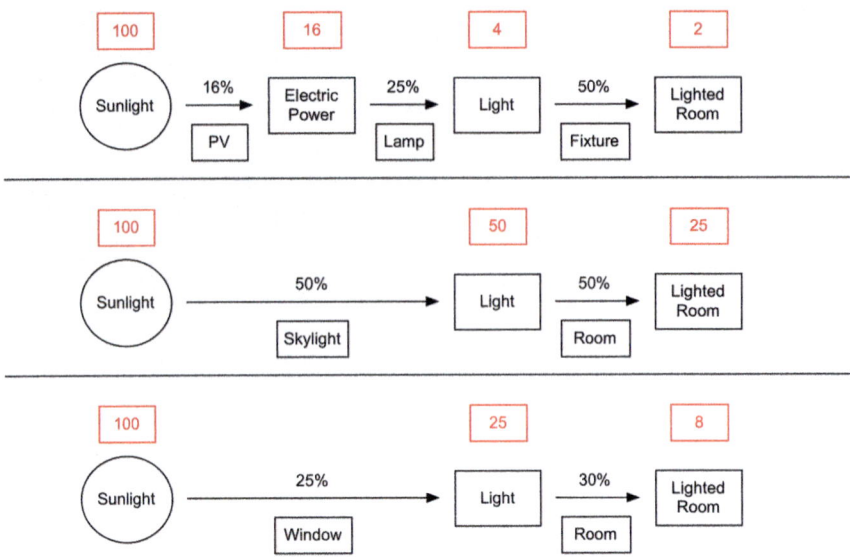

Fig. 3.33 Electric conversion versus direct use of sunlight for room lighting

3.3   From Static to Dynamic Systems

As the architecture, engineering and construction industries shift towards pursuing low and ZNE design strategies as standard practice, it is anticipated that design teams will increasingly explore the integration of dynamic, environmentally responsive facade technologies. While static facade systems serve as a practical option for lighting and HVAC energy reduction efforts, the resulting indoor environmental conditions are often unacceptable to occupants due to the inability of static systems to respond to daily or seasonal changes in sun and sky conditions, to manage air flow, or effectively manage between the dynamic range of outdoor solar and lighting conditions and the occupants' desired dynamic range indoors. As a result, static facades that "optimize daylight" through maximizing physical transparency often lead to retrofits and occupant modifications over the project life cycle to address glare and solar over-heating which, in turn, serve to greatly reduce the anticipated energy savings and IEQ benefits. Alternatively, static facades that incorporate extensive shading, small window apertures, and glazing technologies to reduce visual transparency fail to achieve energy (e.g. ASHRAE 90.1-2013) or IEQ (e.g. LEED EQ) objectives.

In concept, dynamic facade systems are capable of continually adjusting the thermal and/or optical properties of the envelope to seek the optimal balance between energy and human-factors objectives for any given exterior environmental condition and interior occupant requirement (Fig. 3.34). For dynamic, integrated facade systems to perform effectively from an occupant perspective requires the development of systems that are locally configurable (i.e. granular) and capable of modulating exterior conditions to deliver the indoor environmental conditions desired by

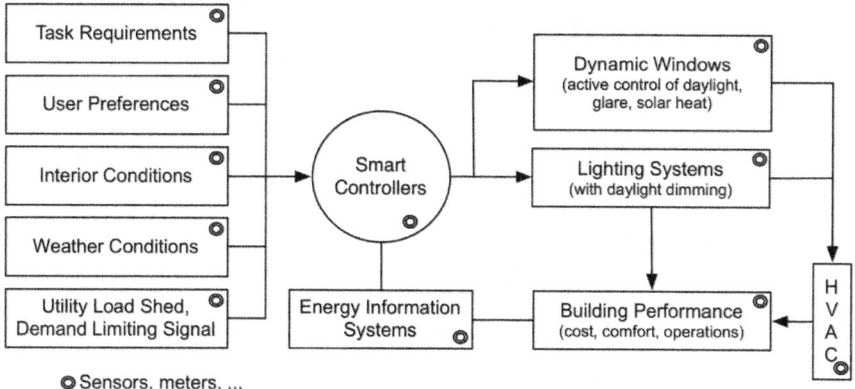

Fig. 3.34 Conceptual diagram of an intelligent control system. Maximizing performance requires full integration with all building systems. *Image credit* LBNL

building occupants. At the extreme, an ideal thermal management solution would minimize or even eliminate the local need for HVAC, and an ideal optical system would minimize or eliminate the need for electric lighting. Effective real-time performance, and performance over the project life cycle requires systems that are context-aware, robust, capable of learning, and interoperable with other building (and grid) systems. These latter topics are discussed in the following section (Sect. 3.4).

3.3.1 Granular Design

Making dynamic systems work in practice requires a granular approach to design. Granular, in this context begins with a recognition of the different needs of occupants carrying out different visual tasks at different locations relative to the window and then refers to the subdivision of facade elements horizontally and vertically across the facade and to the assignment and distribution of various environmental functions (e.g. solar and glare control) to different facade layers, allowing the geometry and material properties of each layer to be developed more specifically for the assigned function. This is further complicated by weather patterns and the dynamic path of the sun hourly over a day and daily over seasons. For dynamic systems, a granular approach enables far greater possibilities for creating personalized facade configurations to suite the preferences of building occupants as conditions change. Granular, layering of the optical control solutions is pragmatically the only way to improve the facade's ability to respond to the daily and seasonal changes in the dynamic range of exterior solar and glare conditions by expanding the possible number of system configurations as well as enabling the system to benefit from the performance of multiple layers in aggregate.

The example presented in Figs. 3.35, 3.36, 3.37 and 3.38 shows a dynamic facade applied as a retrofit in 2005 to the Kreditanstalt fur Wiederaufbau (KfW) building

Fig. 3.35 Dynamic facade applied as a retrofit in 2005 to the Kreditanstalt fur Wiederaufbau (KfW) building located in Frankfurt Germany, originally built in 1968. The facade incorporates dual-pivot manually operable windows for natural ventilation, which are dynamically shaded by automated exterior sunshades that retract into the spandrel zone of the facade when not required

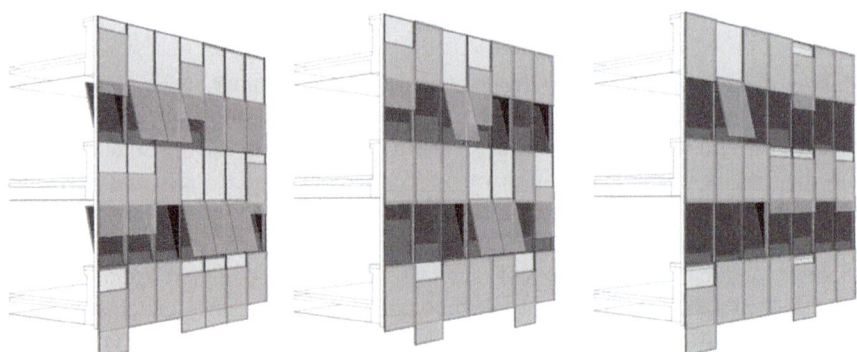

Fig. 3.36 Three potential configurations of the KfW facade system illustrating general variations in exterior shading of windows from greatest (*left*) to least (*right*). Note each example also includes local variations. Image drawn by Sue Long Lee

located in Frankfurt Germany, originally built in 1968. The facade incorporates dual-pivot manually operable windows for natural ventilation, which are dynamically shaded by automated exterior sunshades that retract into the spandrel zone of the facade when not required. The sunshades consist of two layers of glass with an

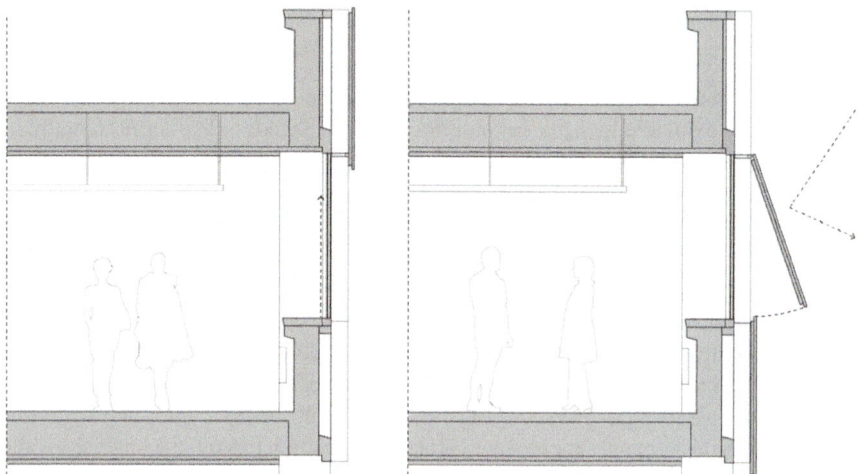

Fig. 3.37 Sections of KfW facade retrofit. Automated exterior sunshades deploy to block direct sun (*right*) and retract into the spandrel zone of the facade when not required (*left*). An interior, bottom-up translucent roller shade provides glare control from the bright sky and the option for visual privacy while permitting daylight from the upper zone of the window. Image drawn by Sue Long Lee

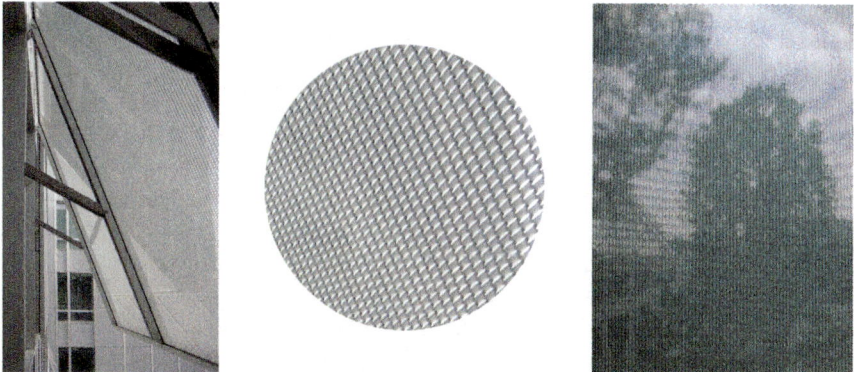

Fig. 3.38 View (*right image*) preserved from building interior looking through deployed sunshade (*left image*) (from a distance of 2 m). The sunshades are a laminated glass unit consisting of two layers of glass with an expanded 3-dimensional metal interlayer (*center image*) that acts as an angular selective screen, blocking direct sun and occupant view of the solar disc while preserving partial view to the exterior when deployed

expanded 3-dimensional metal interlayer that acts as an angular selective screen. An interior, bottom-up translucent roller shade provides glare control while permitting daylight and allowing the occupant to maintain an unobstructed view to the outdoors. In addition to manual control over the operable windows and interior shading devices, occupants have control over the positioning of the exterior sun shades via a wall control integrated with the room electrical lighting controls.

3.3.2   Dynamic "Smart" Glazings

After more than 20 years of development, mature dynamic glazings are now commercially available at sizes suitable for building facades. The term "dynamic," in this context means capable of changing to various optical transmission states, where each state has different performance characteristics with respect to solar gain and daylight control. Dynamic glazings include several different technologies, which can be categorized as either controllable or non-controllable. Controllable (or "active") technologies change their state on demand, in response to signal input from Building Automation Systems (BAS), networked sensors, or occupant activated wall controls. Active technologies include liquid crystal, suspended particle, and electrochromics. Non-controllable (or "passive") technologies change their state in response to input from the local environment. For example, photochromic glazing materials change their transparency in response to light intensity, and thermochromic materials modulate the amount of transmitted light in response to changes in the temperature of the glass which is impacted by solar gain and air temperature. The primary limitation of "passive" behavior in dynamic systems is that building lighting and space conditioning needs rarely correlate directly with simple environmental variables. And, occupants desire the ability to create a clear window view on demand, even if this action is rarely taken. Both thermochromic (temperature activated) and electrochromic (electrically switched) are examples of commercially available dynamic glazings, although both are still in the early stages of market acceptance. In addition to the commercial products now available, new nanotechnology based solutions for electrochromics that promise even better performance at lower cost are emerging from R&D labs (Granqvist 2014; DeForest et al. 2015) and work continues on thermochromic and photochromic solutions. As discussed in the following section (Sect. 3.4), active dynamic glazings must be linked to real-time sensing infrastructure and building automation systems that account for occupant needs and preferences to capture their full potential.

Switchable variable-tint electrochromic (EC) windows entered the market in 2006 and are now in production in the U.S. by multiple vendors at high-volume manufacturing plants, enabling lower cost and larger area window products to be specified (Lee et al. 2013). Electrochromic glazings are a controllable technology developed with the purpose of modulating solar radiation to control solar heat gains and glare by transitioning from a clear to darkened state on demand with an applied low voltage signal, while preserving a clear, but colored window view. The leading commercial electrochromic coatings today are applied to float glass using sputter deposition processes, which are similar to the coating process used in the manufacture of low-e glass. The coating consists of at least five primary functional layers (Fig. 3.39) including two transparent conductors that apply the switching voltage, which in total are less than 1/50th the thickness of a human hair (4 μm). After the EC coating is applied, the EC coated glass layer must be combined with different glazing substrates, low-e coatings, gas fills, and spacers and fabricated into standard IGUs. Since current generation coatings are absorptive, they are general positioned on the #2 surface (inner surface of outer glass) and there must be a low-E coating

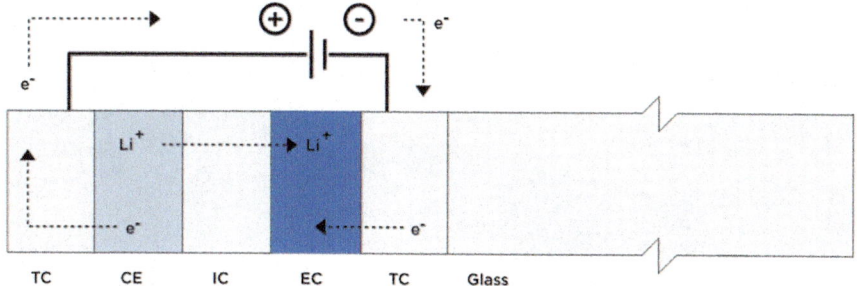

Fig. 3.39 The transparent conductor (TC) layers form a sandwich around the electrochromic (EC) layer, the ion conductor (IC) and the counter electrode (CE). *Image credit* Sage Glass, LLC

Fig. 3.40 Transition in EC tint state from fully clear (*left*), to fully darkened (*right*). Optical properties for each state are (Tint 1: VLT = 0.58, SHGC = 0.46; Tint 2: VLT = 0.40, SHGC = 0.29; Tint 3: VLT = 0.20, SHGC = 0.16, Tint 4: VLT = 0.03, SHGC = 0.09). *Image credit* View Glass, LLC

either as the top layer of the EC or on the facing piece of glass so as reject the absorbed energy to the exterior.

One of the primary benefits of EC technology is that the glazing can be modulated to intermediate states between clear and fully tinted as shown in Fig. 3.40. While the coatings can technically be controlled to reach any intermediate state between their maximum (typically Tv = 0.60, SHGC = 0.40) and minimum states (typically Tv = 0.01 and SHGC = 0.09) for purposes of control uniformity suppliers typically over 4 standard states: a max and min and two intermediate levels. Optical properties for each state are indicated in Fig. 3.40. As shown in Fig. 3.39, the transparent conductor (TC) layers form a sandwich around the electrochromic (EC) layer, the ion conductor (IC) and the counter electrode (CE). The glass is darkened by applying a positive voltage to the TC in contact with the CE which

Fig. 3.41 Facade glazed with EC units subdivided vertically. Tint state of horizontal groups can be independently adjusted to improve glare control while preserving clear views to the outdoors. *Image Credit* Sage Glass, LLC

causes lithium ions to be driven across the IC and inserted into the EC layer, while a charge compensating electron is extracted from the CE, flows around the external circuit, and is inserted into the EC layer (Sage 2016[2]). The darker the tinted state, the more solar radiation and glare are reduced. Reversing the voltage polarity causes the ions and electrons to return to their original layer, causing the glass to return to a clear state.

The tint state of horizontal groups can be independently adjusted to improve glare control while preserving clear views to the outdoors (Fig. 3.41) if the controls structure is sufficiently granular. This can be achieved with separate pieces of glass or by introducing additional electric busbars in a single piece of glass. In addition to vertical windows, ECs can also be used for skylight applications (Fig. 3.42).

When integrated within a double-pane IGU, typical EC windows have an upper visible transmittance range of ~0.30–0.60 and a lower range of 0.01–0.1, depending in part on the properties of the various substrates and the SHGC ranges from 0.07 (fully tinted state) to 0.40 (clear state). For example, Fig. 3.43 presents a plot of various Tvis and SHGC values for an example EC double-pane IGU configuration. The Light-to-Solar-Gain (LSG) ratio is found to be relatively constant.

---

[2]http://sageglass.dreamhosters.com/technology/how-it-works/.

Fig. 3.42 Image of 255 m² ceiling skylight of the Connor Group headquarters in Dayton Ohio, glazed with View Glass, LLC. The EC glass switches from clear (as shown) through several tint levels based on the sun's position in the sky. *Image Credit* View Glass, LLC

Light-to-Solar-Gain is calculated as the ratio of the visible light transmittance (Tvis) to the Solar Heat Gain Coefficient (SHGC). Four tint states are plotted, from tint 1 = clear state, to tint 4 = fully tinted state. For comparison, three different static window configurations (A, B, and C) are included. Note that the "clear single pane" option is presented for reference only, and is not a realistic alternative.

Electrochromic coatings are providing new sun and glare control capabilities not previously available as commercial products. The potential benefits of effective application of EC glazings are wide-ranging include increased provision of views, and improved visual comfort and thermal comfort for occupants, and new energy management capabilities for building owners including significant electrical lighting (when integrated with a dimmable electric lighting system) and cooling energy savings, reductions in HVAC capacity and peak demand, and the potential to participate in active load management, demand side management and demand response. However, market adoption remains at a slow pace as with many other new introduced products. The cost and complexity of the controls integration remains an obstacle for some owners, and the transition speed is slower than some would like with large sheets of glass. While the switching lifetime has been proven in ASTM tests many owners want to see examples that have been in place for 5 + years. As noted by Lee et al. (2013), publically available studies of the technologies in occupied buildings (with validation by independent third parties) are

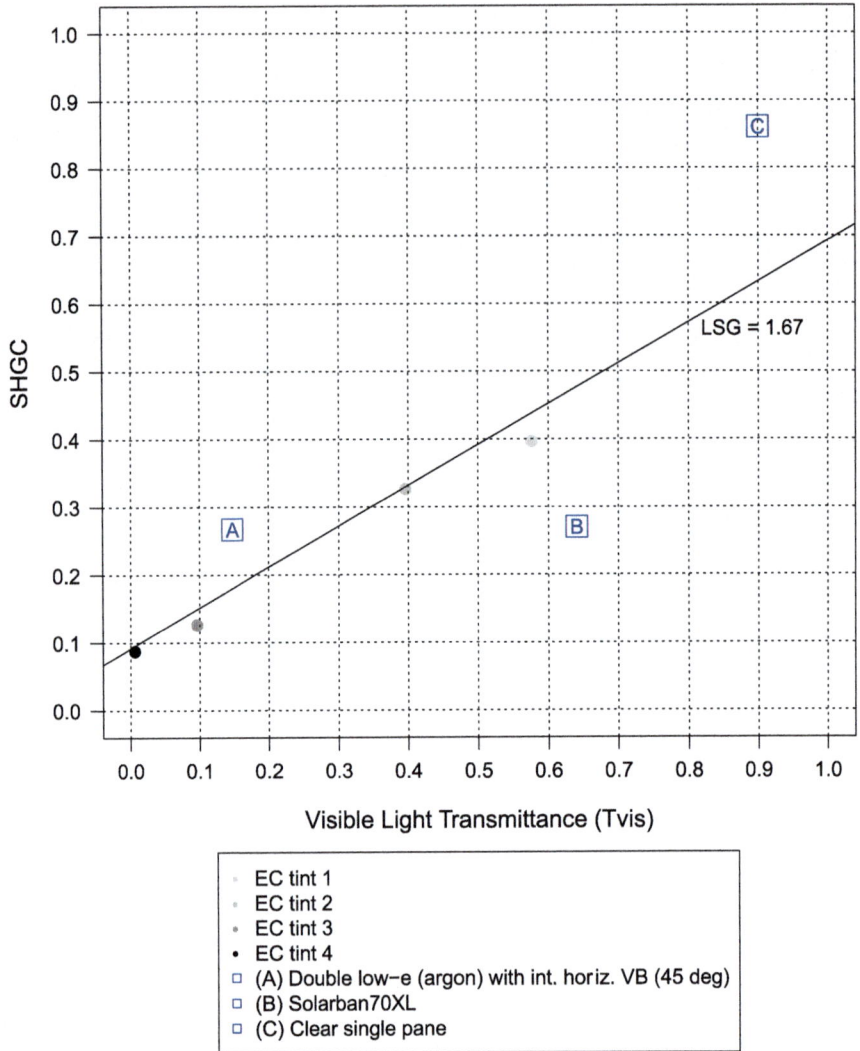

Fig. 3.43 Comparison of optical properties of an example EC double-pane IGU with three alternative static window configurations (*A*, *B*, and *C*). EC data was calculated by LBNL Window 7.3 and was obtained from View Glass (2016). The example EC used is a double-pane IGU configured with outboard lite = 6 mm clear FT with EC coating on #2, inboard lite = 6 mm clear, cavity = 12.7 mm, gas fill = 90% argon

extremely limited. While ECs cannot completely block view of the solar disc, for most practical applications the current low end Tv of 0.01 appears to be adequate for glare control. Some of these challenges are currently being addressed with continuing improvements in the current technologies and some emerging next generation electrochromics technologies. One example is dual band near infrared switching electrochromic (NEC) glazing, which initially remains transparent to visible light as the near IR is reduced, then can be further darkened in the visible if more control is needed (Deforest et al. 2015). Even as the coatings improve the "systems integration" challenges of linking active devices to sensor and control networks remains to be further improved, as discussed in Sect. 3.4.

3.3.3   Dynamic Light Redirecting Systems

Dynamic light redirecting systems offer the potential to better-address one of the central challenges associated with daylighting multi-story buildings, which is the delivery of sufficient daylight beyond the nominal perimeter zone (i.e. >5 m from the facade). The SunCentral System™ (SunCentral 2016) is designed to autonomously track the sun at roof level (Figs. 3.44 and 3.45) and redirect sunlight as a static collimated beam to optics integrated into an exterior horizontal shade element at each floor level (Fig. 3.46). The beam is then concentrated and directed into a

Fig. 3.44 Exterior view of the SunCentral *SunBeamer*™ installed in a skylight configuration. *Image credit* SunCentral

Fig. 3.45 Interior view of the SunCentral *SunBeamer*™ installed in a skylight configuration. *Image credit* SunCentral

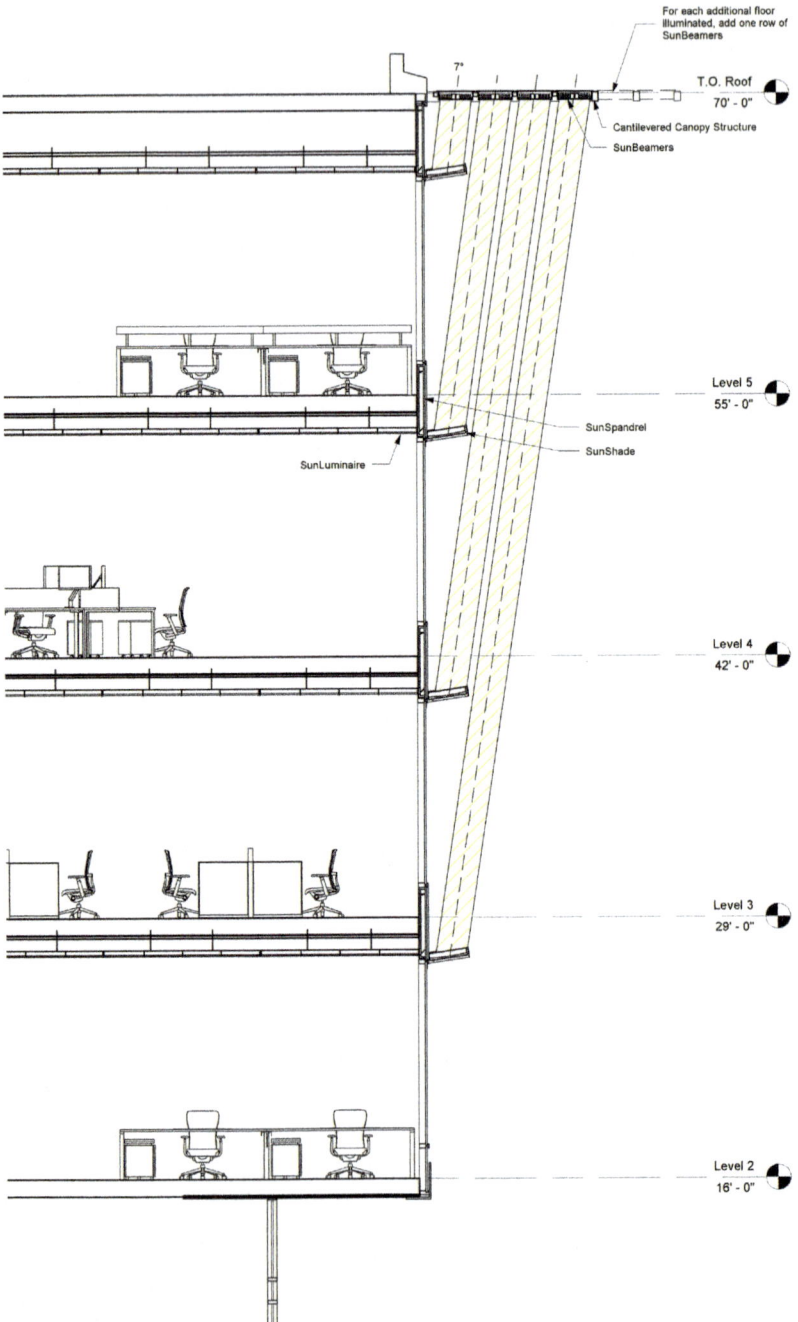

Fig. 3.46 Section view showing the SunCentral System™ installed in a facade configuration. The technology tracks the sun autonomously, projecting a stationary collimated beam of sunlight along any side of the building, including the north side, and can also be installed to beam through an atrium or light well. *Image credit* Sun Central

spandrel element, which channels the sunlight into luminaires within the building (Figs. 3.47 and 3.48). The luminaires are designed with integrated LED lighting and photocontrols to automatically modulate the output of electrical lighting in response to available daylight. The technology offers the potential to both extend

Fig. 3.47 Rendered (conceptual) view of building facade and interior ceilings showing daylight delivered from SunLuminaire™ fixtures. *Image credit* SunCentral

Fig. 3.48 Interior view of actual installation of SunLuminaire™ fixtures. *Image credit* SunCentral

the useful daylit area of large multi-story buildings as well as extend the useful hours of operation on a daily basis. As a technology dependent on collimated beam sunlight, effectiveness may be limited by climates with significant hours of non-clear sky conditions, or sites with significant hours of overshadowing. Given the "centralized" nature of the technology these systems are also best at delivering light over large floor plates with constant high occupancy- they are not particularly well suited to provide light only in limited occupied zones for limited periods of time.

3.4   From Integrated to Interconnected Systems: Internet-of-Things-Enabled Perimeter Systems

Active use of the building envelope (e.g. solar control, daylighting, natural venti-lation, and charging/discharging thermal mass, energy harvesting) paired with controllable lighting and HVAC systems is a complex design challenge. However, driven in part by typical building codes, application of building technology often focuses on the efficiency of individual components rather than consideration of the overall performance of multiple components working as a system. This fragmented approach needs to shift to an integrated, context-aware dynamic perspective that addresses the facade as a system that is responsive to "performance needs" at three different levels: (1) comfort and task performance needs of the occupants; (2) en-ergy and economic needs of the building operator; and (3) the local or regional needs of the utility grid.

While significant effort has been placed on "integrated design" practices that seek to achieve greater levels of system integration during the design stage, the operational performance of integrated systems in the occupied building is limited by a number of barriers. These include (1) the lack of interoperability between various technologies, (2) challenges in deploying and maintaining large sensor arrays (e.g. unit cost, commissioning, calibration), (3) lack of detailed, granular, contextual data to drive effective real-time operation, (4) poor or non-existent mechanisms for fault detection and diagnostics, (5) lack of occupant feedback to validate controls assumptions or make adjustments, (6) lack of holistic controls optimization frameworks (due in large part to #5). From a process point of view, design concepts may not be adequately conveyed to and implemented by the construction team, and the hand off to facility managers and occupants is often incomplete and imperfect. Improvements and innovations are limited by, (6) the lack of frameworks for systems to gather and interpret performance data and learn over time, and (7) the lack of a mechanism to store and share knowledge across projects and design team members.

The result of these limitations has been failures in building performance and a resultant aversion among building designers and contractors to adopt complex but promising technologies in favor of "simple" control strategies based on the cautionary view that "simple is usually better." Entirely passive, fixed solutions seem unlikely to properly address the wide range in climate and user needs. Asking occupants to become de facto facility managers and adjust light levels, blinds, thermostats etc. seems equally unlikely. However, fully automated systems risk alienating occupants when they fail to deliver desired comfort conditions. The real world perspective also suggests that occupants may adjust building features for comfort, but will not reliably manage energy performance objectives. We suggest it is time to challenge the common knowledge that "complex controls will never work" and that hybrid models cannot be adapted to support local occupant needs.

The sensors and controls industry globally is now in the midst of a revolutionary change driven in part by the rapid advance of the "Internet of Things" (IoT) movement. The Internet of Things is the network of physical objects— devices, vehicles, appliances and other items embedded with electronics, and sensors, and linked by software-based network connectivity—that enables these objects to collect and exchange data.[3] IoT is based on four critical elements: (1) low cost, distributed powerful sensors and embedded computing, (2) wireless communications; (3) cloud based data storage and computation, and (4) shared interoperable protocols. Much of this technology and infrastructure was created and driven initially by the smart phone industry, but is rapidly gaining traction in numerous other business realms including the building industry where the LED revolution in the lighting community is leading the way (TCLA 2016[4]). It will likely be further accelerated by massive RD&D investments underway to develop autonomous vehicles where distributed sensing and controls—well beyond the needs of a dynamic building envelope—will need to be developed and perfected and manufactured in volume. While the technology involved is powerful and fascinating, perhaps more important are the trends in cost and functionality. A recent survey of costs of key sensors and communications chips indicates that Wi-Fi, Bluetooth, GPS, accelerometers, cameras, and temperature sensors are all available in volume at Original Equipment Manufacturer (OEM) costs of ∼US$1 each. The availability of these powerful, low-cost building blocks suggests that there should not be a large cost penalty to adding this infrastructure to smart facades.

Figure 3.49 presents a conceptual framework for the design of Internet-of-Things-enabled Perimeter Systems (IoTePS). The IoT movement can be leveraged within the building design domain to develop context-aware, interoperable building components that work to optimize the comfort and resource efficiency of buildings throughout the project operational life-cycle. The IoTePS framework is conceived as a vehicle to explore how the ubiquity of sensing and real-time data will transform existing approaches towards building facade and perimeter zone technologies and

---

[3]http://www.itu.int/en/ITU-T/gsi/iot/Pages/default.aspx.
[4]http://www.theconnectedlightingalliance.org/home/.

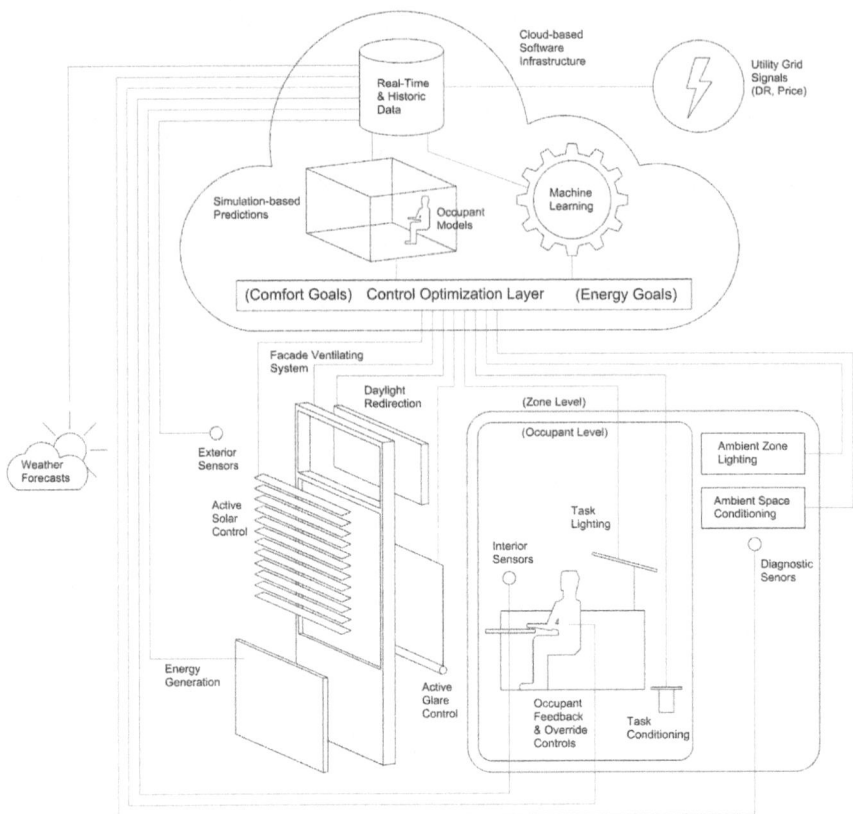

Fig. 3.49 A conceptual framework for the design of internet-of-things-enabled perimeter systems (IoTePS)

the performance roles those technologies are asked to play in buildings. Of specific interest is the transformation of the building facade from a sealed and static element to a dynamic filter, operating in real time to manage a range of grid-level, building-level, and occupant-level performance goals. Charting the functional potential of dynamic behavior, informed through detailed real-time and historic sensor and occupant feedback data, will in turn serve as a basis to explore and develop new specific architectural fenestration strategies, both technologies and design approaches, to best meet this potential.

The systematic collection of reliable, detailed, real-time data has the potential to serve as a platform for a number of innovative design and control strategies. For example, a simulation model of the physical space, informed by past outcomes and weather forecasts can be used to predict changes in energy and IEQ variables to determine the optimal control scenario to minimize energy consumption while maintaining (or improving) IEQ outcomes, with the potential of occupant-level resolution if task lighting and conditioning systems are in place. Occupant feedback, obtained through event-based prompts or monitoring of occupant adjustments and overrides over time can be analyzed using machine learning techniques in context with concurrent physical sensor data to generate and continually refine personalized models of occupant comfort. These models can, in turn, be used to determine dynamic setpoint controls for both task-level and ambient lighting and space conditioning systems to minimize overheating and overcooling, or any arbitrary control variable of interest (e.g. acceptable levels of view occlusion to outdoors, vertical daylight illuminance at eye level, luminance contrasts in the field of view etc.).

The "smart grid" is the subject of much discussion and investment vis a vis the role of renewables and storage, and the decarbonisation of conventional generation sources. In the U.S. almost 75% of the electric grid powers buildings (Buildings Energy Data Book 2011) so it is important to look carefully at the role of the building in creating and managing electric loads as well generating power. At the grid-level, once smart and responsive systems are in place to manage comfort and energy in buildings, the building can play a more complex interactive role with the electric grid and the potentials for effective onsite power generation and storage are increased (Lee et al. 2015). The concept of "transactive energy" (GRIDWISE 2016) envisions that each electric load can communicate with any load in any building and negotiate, based on price, which loads will be reduced, by how much and with or without change in service level, to balance grid needs in exchange for a financial transaction. All of this of course must happen real-time, rapidly and invisibly without occupant intervention, once the criteria are set. The future smart electric grid will be dynamic and automated with buildings fully engaged, and ideally with performance goals set by building occupants and operators. Smart active facades could provide great value in such a system, given their crucial role impact on thermal and lighting systems in buildings.

3.5 From Closed-Loop to Human-in-the-Loop Systems: Incorporating Human Factors Models and Feedback From Real Buildings in Use

As noted previously, manual and automated facade shading and daylight dimming control systems have been commercially available for decades but they have never had widespread impact due to design and operational complexity and costs. The current challenge is to develop systems that are capable of routinely delivering acceptable (or preferred) environmental conditions to occupants over an annual range of environmental conditions while effectively implementing low energy strategies that may involve demand response, charging/discharging of thermal mass, natural ventilation, and integrated electrical lighting and HVAC energy optimization. Meeting this challenge requires understanding both the complexity of contemporary facade systems and the appropriate dynamic building physics as well as the complexity of end-user needs and preferences.

While windows are often conceptualized as apertures that are either configured in a fully open or fully shaded state, the reality of facade shading configurations is more complex. Observational studies show that different people prefer significantly different facade configurations under the same sun and sky conditions, motivated by a range of factors beyond the domains of energy and lighting, such as privacy, task performance and ease of operation. In addition, the desired rate of change for adjustments varies among occupants (e.g. instantaneous, hourly, daily, seasonally, never). Figure 3.50 illustrates the complexity of the challenge for contemporary facade designs simply in terms of interior operable shading.

Figure 3.50 is a composite of interior elevations created from observation of the southeast facade sections of the San Francisco Federal Building. To systematically examine how patterns of shade use impacted daylight availability and views to the outdoors, facade sections were photographed in HDR at 5-min intervals each workday and composited to visualize shade configurations over a 5-week period during the summer. The number shown in brackets below each facade section indicates the time-and-area-weighted average of glazing covered by interior shades (discounting the lower row of windows (in grey) which were generally occluded by office furniture). The numbers in green indicate the total number of shade operations observed for each window over the 5-week period. The figure shows that the majority of occupants shaded over half (50%) of available facade glazing above desk height, and rarely or never adjusted the shade position (note the absence of a green number indicates no shade adjustments were observed). Notably, nearly all participants maintained a small, unshaded area at or below seated eye level to preserve an unobstructed view to the outdoors. Occupants who did not (or rarely) lower shades were often found to have made personal modifications to their work area or adapted their behavior to accommodate direct sun and glare conditions. These modifications are discussed in more detail in Chap. 6 but the primary lessons learned are that (1) people do not actively manage their shades but rather leave them in a fixed position that excludes sunlight, but (2) they continue to desire at least limited view access.

Effective operation of automated systems requires that control system assumptions are validated and refined against empirical models of occupant behavior and subjective preferences in order to ensure long-term user acceptance. In addition, convenient user interfaces are needed to enable user-overrides and to solicit subjective data to validate and refine control assumptions. To address the first concern, a data set was developed for analysis by coding each observed shade operation (from the above example) with simultaneous physical measures of exterior and transmitted global vertical illuminance and irradiance, Mean Radiant Temperature (MRT), approximated from a globe thermometer adjacent to each participant, and multiple luminance-based measures and glare metrics. This data set was used as a basis to examine alternative predictive models of shade operation. Using single-variable logistic regression as a tool to model behavior, roller shade

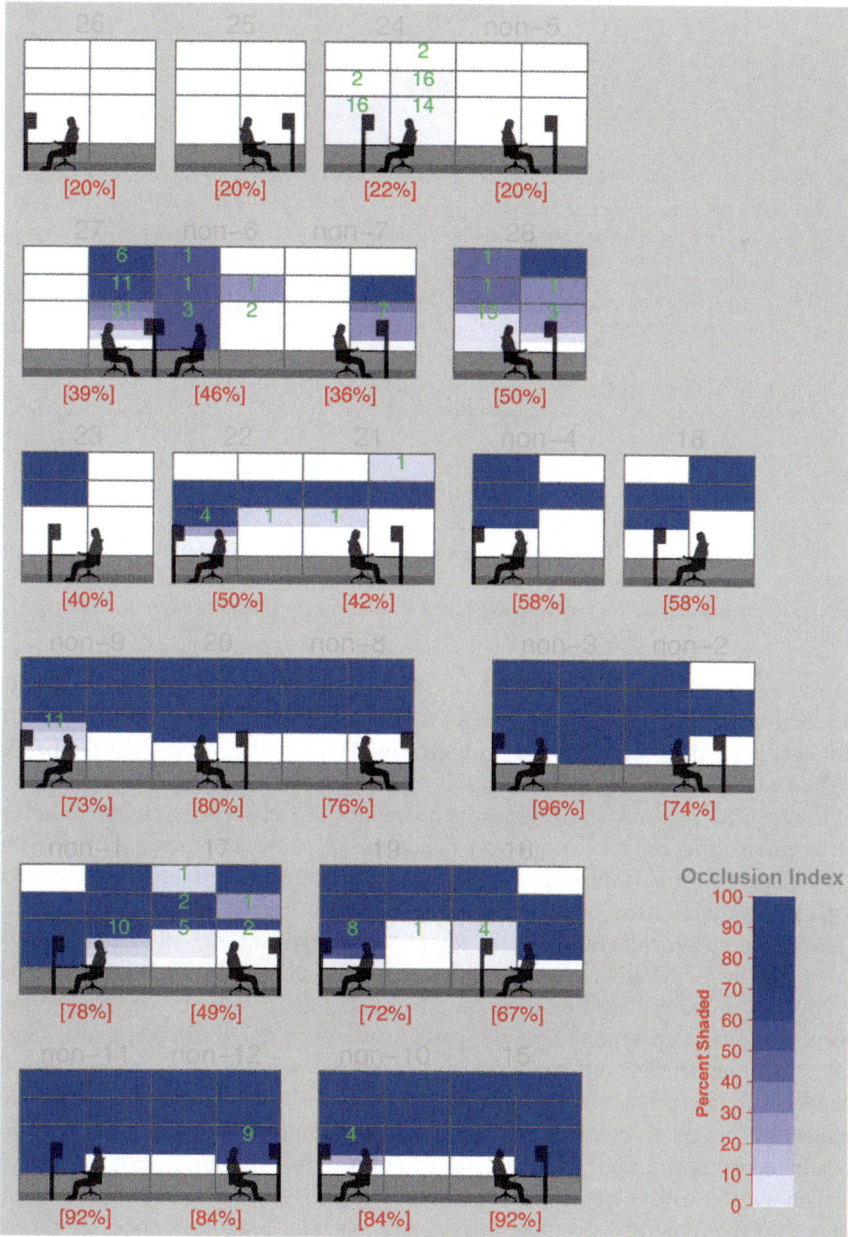

Fig. 3.50 Summary composite of shade configurations and use behaviors over five-week monitoring interval, ordered from top to bottom by low to high occlusion level. In total, 245 shade operations were observed from (N = 14) participants over 5 weeks (25 workdays), leading to an average of less than 1 (0.7) shade operation per person per day

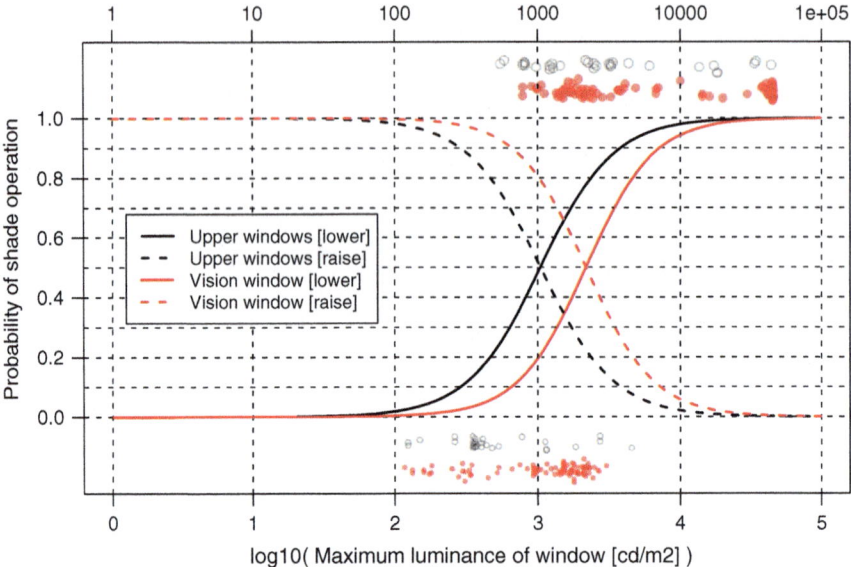

Fig. 3.51 Empirically derived logistic regression models for maximum window luminance

operations were found to be related to the stimulus intensity of a number of interior physical variables, where the probability of a shade lowering event increased with stimulus intensity. Measures of maximum window luminance were found to be the highest ranked predictors for both upper and lower shade groups.

Figure 3.51 presents an example shade control model showing the relationship between shade control behavior and maximum window luminance. Overall, the models showed high probabilities for shade deployment at stimulus levels below control thresholds used by existing occupant shade control models. All existing indicator thresholds examined were found to overestimate the stimulus intensity associated with the lowering of shades, possibly leading to overestimations of daylight availability based on even the most "optimistic" assumptions for the frequency of shade operation. Notably, participants were found to operate the upper daylight zone shades differently than the vision window shades. Participants showed a higher probability for lowering the upper shades for any given stimulus intensity. This result contradicts common design guidance when subdividing the facade to include an upper daylight zone (and increasing floor-to-ceiling-height).

The ability to develop innovative energy efficient fenestration systems, implement effective controls, diagnose faults, and maintain long-term end-user acceptance requires empirical understanding of the environmental conditions acceptable to (and preferred) by occupants. Without consideration of these conditions, functional requirements cannot be established for individual components, controls cannot be optimized, and guidance cannot be established for managing performance trade-offs during design. Data-driven models such as those presented in Fig. 3.51

serve as a mechanism for control systems to learn the preferences of individual building occupants to continually adjust and refine operations.

To improve fidelity between control assumptions and the environmental conditions desired by occupants, it is critical for robust models of occupant behavior to become an integral part of dynamic control algorithms. It is equally critical that the models be informed through observation and evaluation of human behavior in buildings in use, that it is correlated to reliable measurement of associated physical conditions and that this information is fed back into the design process, not simply to improve model fidelity, but to drive innovation in the design and operation of integrated facade systems. The topic of field evaluation and feedback is discussed in detail along with emerging occupant-aware monitoring approaches in Chap. 6. Beyond simply a controls challenge, integration of human factors considerations presents a framework for exploring refinement of facade systems and individual components to deliver more personalized environmental conditions.

3.5.1 Granular Sensing for Personalized Control: Utilizing Image Based Lighting for Viewpoint Specific Dynamic Glare Control

The growing desire for more granular, dynamic facade systems with high levels of occupant comfort requires reevaluation and improvement over existing approaches for environmental sensing. In most conventional applications, dynamic systems are controlled in an open-loop configuration with a single light sensor located on the building exterior, or in response to a simple schedule derived from solar position. While these approaches may be sufficient to manage solar heat gains, they are incapable of sensing and identifying internal viewpoint-specific visual conditions that often lead to visual discomfort and dissatisfaction with dynamic systems. In such cases, systems are typically re-configured over time by occupants to increasingly occluded positions that unnecessarily overshade the facade to minimize glare at the expense of daylight transmission and views to the outdoors. Alternatively, efforts to physically instrument indoor spaces to capture all possible viewpoints are impractical due to the large number of occupant viewpoint locations, the complexity of sensing required to assess glare (e.g. image-based), and the inability to ideally locate sensors at viewpoint origins due to the presence of occupants and furnishings.

The application of Image Based Lighting (IBL) presents a promising and practical approach to improve daylighting and glare control by tailoring the operation of dynamic systems to specific occupant locations and preferred visual conditions while also improving the overall level of awareness of a dynamic system to its environmental context. Originally developed for cinematic arts applications, IBL uses a calibrated HDR image of the physical environment as a light source to illuminate a virtual object or scene (Debevec 1998; Inanici 2010). While originally intended to make digital objects appear more realistic and appropriately situated in their luminous environment, IBL can be easily applied to simulate a realistic view of the environmental light source from any arbitrary location within a building.

For daylighting applications, the primary sources of glare are typically produced from the sun and sky (e.g. cloud conditions) and modified by the building interior and facade elements. Therefore, if given the luminance conditions of the sky at a specific point in time, the potential for glare can be readily assessed for any arbitrary view from within the building by rendering the viewpoint in Radiance using the sky as an image based light source.[5] Figure 3.52 presents a schematic view of this approach. Each rendered view results in a luminance map that can then be analyzed to assess the probability of glare. Results can, in turn, be used to inform appropriate changes to the facade and interior lighting system. A conceptual application of this approach is demonstrated in the following pages.

The example shown in Fig. 3.52 utilizes a hemispherical HDR image of the sky acquired from the Lawrence Berkeley National Laboratory's SkyCam (LBNL 2016) to render a southeast (SE) facing view from a Radiance model similar to the San Francisco Federal Building's SE facade.

For the purposes of this example, only a single view is shown, however it is possible to simulate an arbitrary number of views for any point in time to assess glare conditions for all occupants who share this open office space.

Previous research by the author has demonstrated the applicability of low-cost HDR sky imaging (e.g. Fig. 3.53) for enabling granular dynamic control of facade systems (Konis 2013). Until recently however, a limiting factor for applying IBL for controls applications has been the lack of a purpose-built sensor platform capable of persistent image acquisition in exterior weather conditions paired with calibrated global irradiance and illuminance measurements.

In 2014, Terrestrial Light, a Berkeley-based startup, developed the HDR sky imaging system (SkyCam) shown in Fig. 3.54. SkyCam takes exposure-bracketed images of the sky at regular intervals (e.g. every five minutes). These images are composited into a single HDR image that records the entire luminance range of the sky. The HDR image is calibrated using global irradiance and illuminance measurements from sensors mounted alongside the skycam (Fig. 3.54). When available at regular intervals, these images can be used to visualize the luminance conditions for a theoretical observer over daily changes in sun and sky conditions. Figure 3.55 illustrates the variation in luminance conditions for the theoretical observer for four points in time on December 29, 2014 (dynamic sky conditions).

Simply the capability to visualize how the building interior and facade systems interact with real skies from the viewpoint of building occupants is extremely valuable for assessing the performance of facade systems. The time-series HDR

---

[5]It is important to note that HDR imaging using CCD cameras is limited by the saturating effect of the solar disc, which is underestimated due to the dynamic range limitations of the CCD sensor. However the solar disc region of an image can be identified in software and the approximate known luminance of the solar disc can be applied to the final luminance map as a post process.

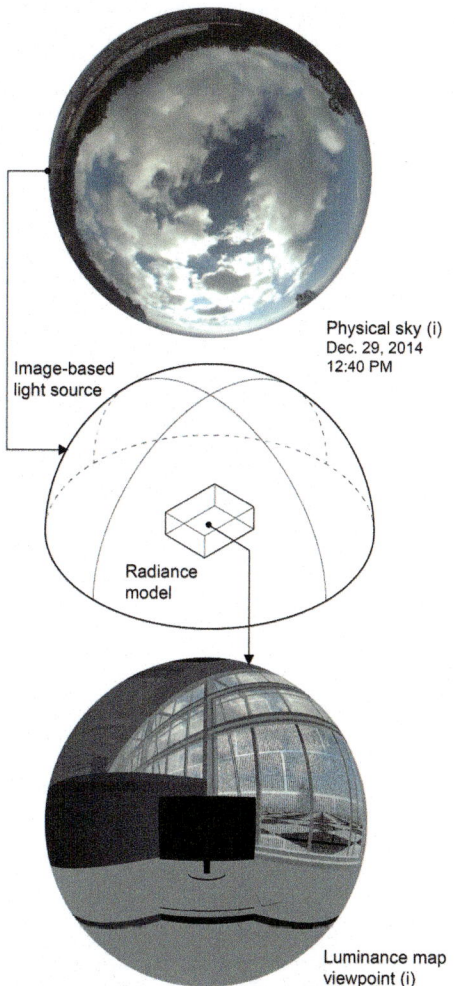

Fig. 3.52 Schematic application of image based lighting combining physical and virtual information to determine individual occupant viewpoint luminance conditions

images acquired from specific viewpoints also can be utilized by control algorithms to make decisions for the state or position of dynamic facade technologies used to control glare. While a range of existing glare metrics are available (e.g. DGI, DGP) to predict glare within the field of view, the response of a glare control system (e.g. shading device, EC) is typically restricted to defined regions of the window wall that can be adjusted in position or state. Consequently, it is more important to assess the potential for glare within specific control regions at the level of granularity enabled by the glare control system. The following example presents a conventional scenario where each window region can be shaded independently by an interior, top-down

Fig. 3.53 Author's own
LDR camera, oriented
vertically with 180°
hemispherical lens

Fig. 3.54 Skycam. *Image
credit* LBNL

fabric roller shade. The scenario assumes that each roller shade is automated and
controlled by an algorithm that requires luminance-based data from each region as a
signal input to inform the position of each shade to deploy only when required for
glare control. Figure 3.56 presents one possible subdivision scheme where regions
of the facade are isolated for analysis using a mask defined and applied as a

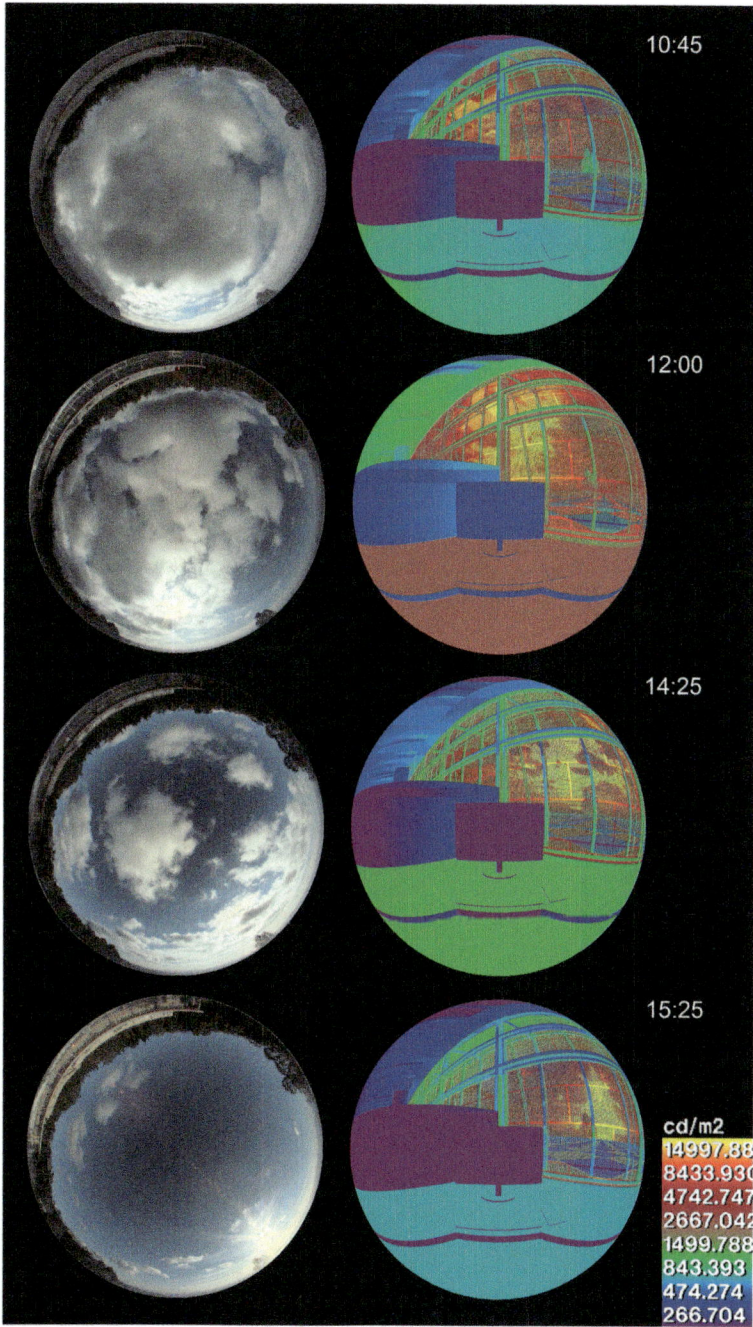

Fig. 3.55 Example of the variation in luminance conditions of the sky (*left*) and for the theoretical observer (*right*) for four points in time on December 29, 2014 (dynamic sky conditions). Images are acquired from a SkyCam located on the *rooftop* of the Lawrence Berkeley National Laboratory's FlexLAB, located in Berkeley, CA

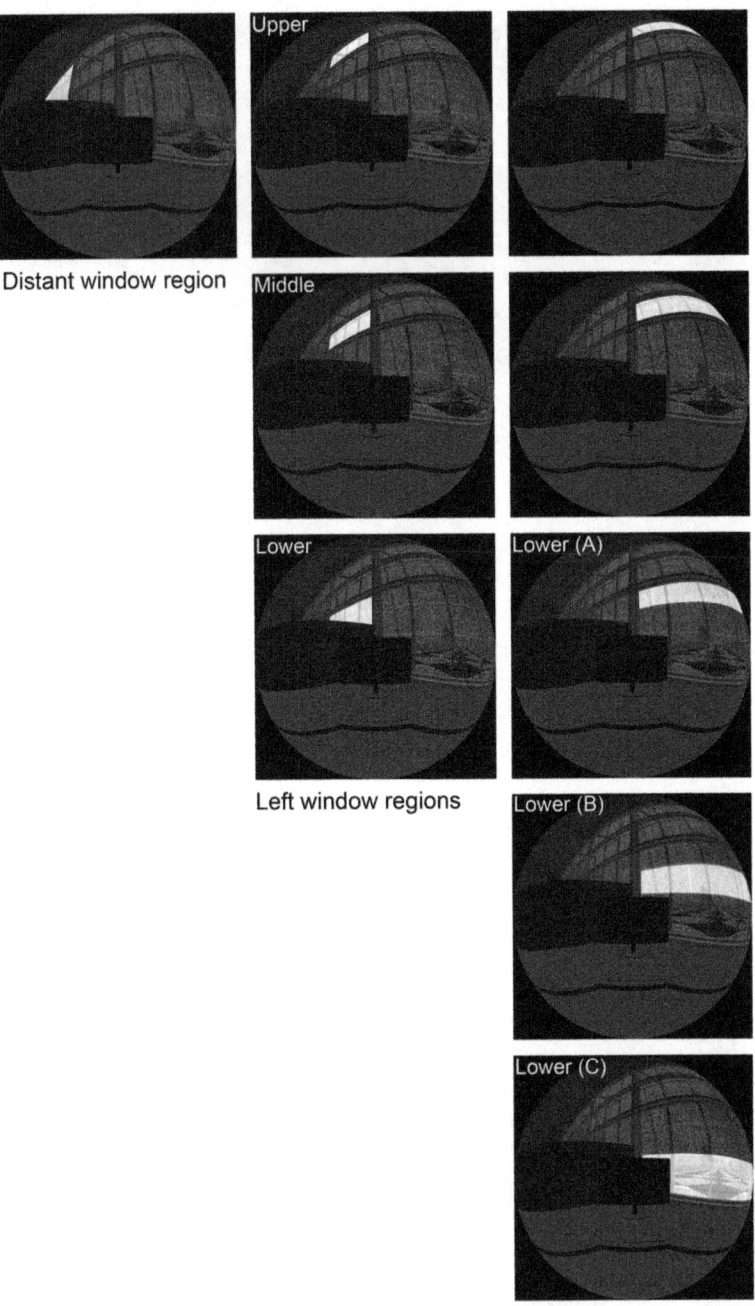

Fig. 3.56 Viewpoint-specific region boundaries defined for image-based window luminance analysis

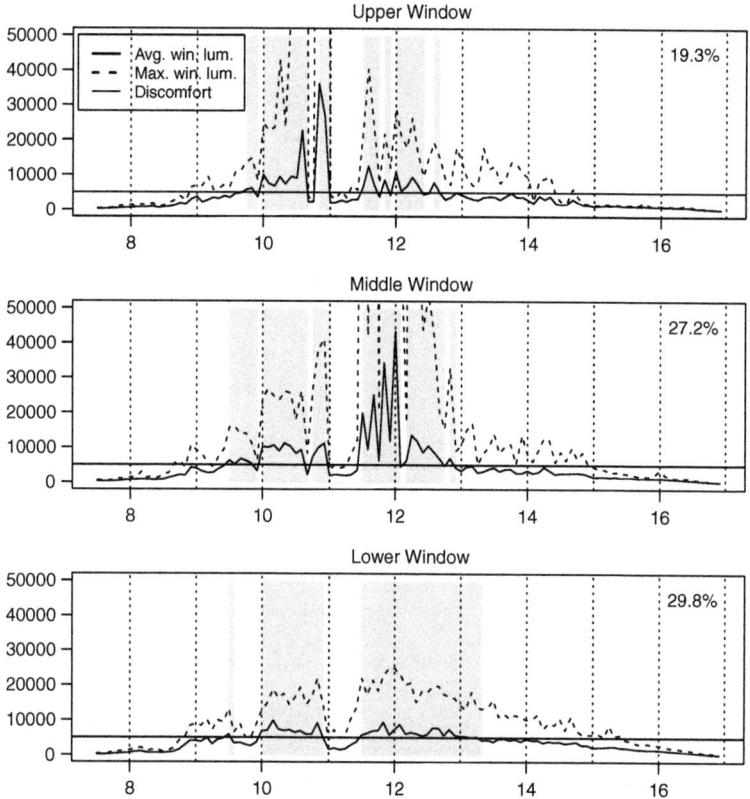

Fig. 3.57 *Left* window regions. Dec 29, 2014. Dynamic sky conditions

post-process to determine the average and maximum window luminance of the region for each point in time. Windows near the observer are defined as one or multiple individual analysis regions, whereas windows in the far field are grouped for simplicity in this example. This simple subdivision scheme leads to nine (9) regions that can be configured independently to optimally balance between the specific observer's personal preference for visual comfort, daylight and view access.

Figures 3.57 and 3.58 present the resulting luminance data for each window region defined for the LEFT and RIGHT windows in Fig. 3.57 based on analysis of the December 29, 2014 dynamic sky data. Y-axis units for luminance are cd/m^2. For each region, measured data is compared to a threshold of 5000 cd/m^2 (horizontal line: average luminance across the region) to determine the shade state control signal (up/down). Periods of the day when the shade is required to be down are shown in grey. The percent of hours (6:00 AM–6:00 PM ST) when the shade is deployed is quantified in the upper right of the figure.

For the LEFT window (Fig. 3.57), the chosen threshold leads to only a small fraction (between 19 and 30%) of the day when shading is required. The results for the

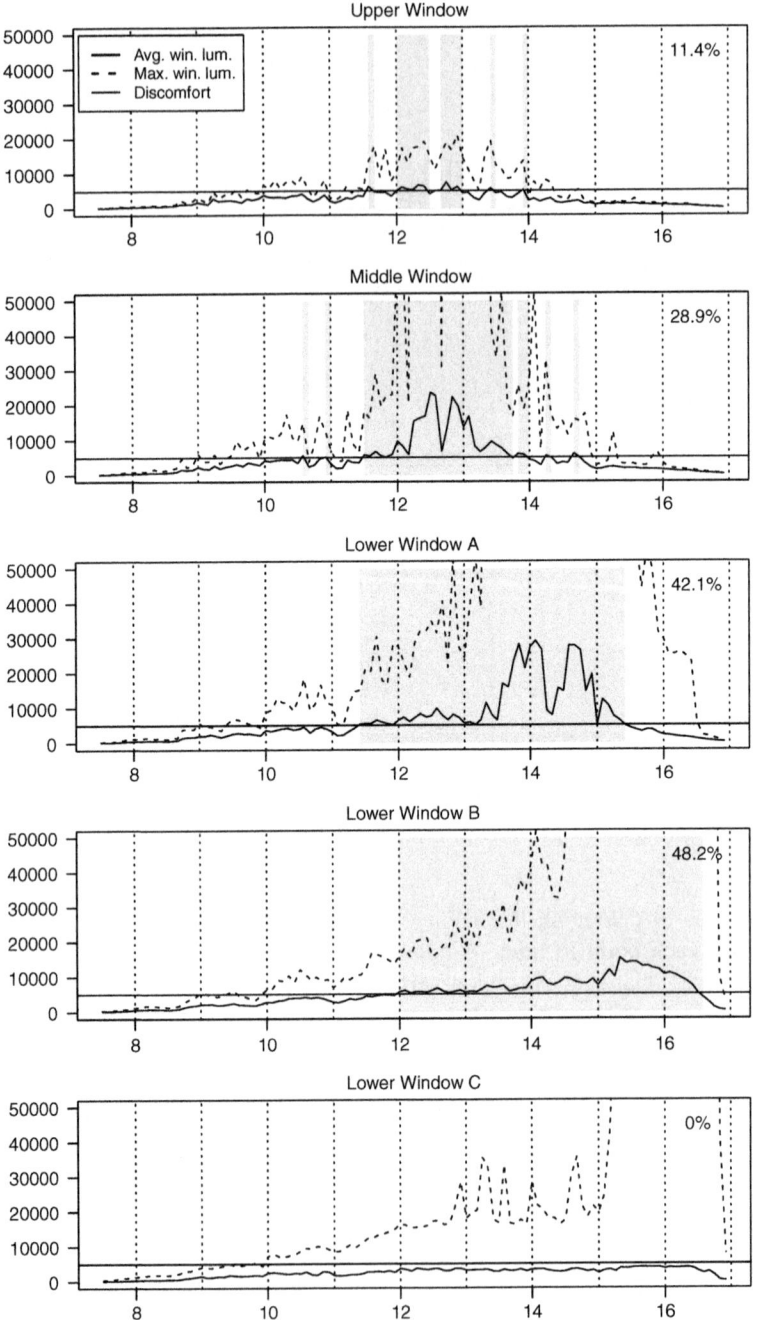

Fig. 3.58 *Right* window regions. Dec 29, 2014. Dynamic sky

LEFT also show relatively similar times during the day when shading is required and indicate long periods in the afternoon when shading is not required. Results for the RIGHT window (Fig. 3.58) show a more complex pattern, where upper window regions can remain unshaded for periods of the day when several regions of the lower window must remain shaded to control glare. Notably, region C of the lower window was found to never require shading, permitting a view to the horizon for the entire day.

In comparison to a conventional control scenario where zoned groupings of shades are deployed or retracted simultaneously, the application of IBL for dynamic control has the potential to improve daylighting performance and occupant comfort by deploying only the shades needed for glare control for specific viewpoints, leading to improved daylight transmission and visual connection to the outdoors. Moreover, by assessing glare directly from the occupant's view, control algorithms can attain a higher level of precision in regard to the luminance conditions actually being experienced by occupants in the zone at any given time. It is important to emphasize here that the present example is a simple demonstration using only one occupant viewpoint. In reality, the various trade-offs between glare, view and daylight availability must be managed for multiple viewpoints simultaneously. However, the value of the approach is in the provision of granular, occupant level exposure data without the need to install a dense grid of physical sensors.

While a wide variety of luminance-based metrics exist for predicating what luminance conditions will lead to glare discomfort, the personalized nature of the control strategy can be readily incorporated into control scenarios that apply machine learning techniques to leverage occupant feedback to determine person-alized predictor variables and threshold values rather than relying on a "standard observer" model. This capability itself presents great potential for improving occupant comfort, due to the ability to adapt controls to specific users and their unique physiological needs and personal preferences.

3.6 Conclusions

This chapter paints a promising but challenging future scenario to provide robust, effective daylighting design solutions that work. New glazing and responsive facade technologies supported by the ubiquitous sensing, control and communications infrastructure of the Internet-of-Things offers the potential to overcome the tech-nical challenges that often limit the capability of existing "integrated" facade and perimeter zone systems to effectively manage a growing array of performance requirements. These requirements include consistently delivering acceptable (or enhanced) environmental conditions for occupants while simultaneously con-tributing to a low-energy building concept and playing an active role in emerging smart grid development strategies. Effective outcomes will also depend on the integration of occupant feedback and new, occupant-aware control methodologies. However these outcomes will not emerge on their own- they will only become feasible and gain traction if adopted, demanded and pursued by many key partic-ipants across the building sector. New design processes, analysis tools, and field-based validation techniques, will help set more rigorous operating expecta-tions for project stakeholders. These performance-based design processes and tools are the subject of the following chapter (Chap. 4). Chapter 6 presents emerging occupant-centered evaluation methods for improving the feedback loop between design expectations and performance in use.

References

American Society of Heating and Refrigeration Engineers (ASHRAE) (2013) ANSI/ASHRAE/ USGBC/IES standard 90.1–2013. American Society of Heating and Refrigeration Engineers, Atlanta

Debevec P (1998) Rendering synthetic objects into real scenes: bridging traditional and image-based graphics with global illumination and high dynamic range photography. In: SIGGRAPH 1998 conference proceedings

DeForest N, Shehabi A, O'Donnell J, Garcia G, Greenblatt JB, Lee ES, Selkowitz SE, Milliron DJ (2015) United States energy and CO_2 savings potentials from deployment of near-infrared electrochromic glazings. Build Environ 89:107–117

Granqvist CG (2014) Electrochromics for smart windows: oxide-based thin films and devices. Thin Solid Films 564:1–38

GRIDWISE (2016) http://www.gridwiseac.org/about/transactive_energy.aspx. Last accessed 29 Aug 2016

Inanici M (2010) Evaluation of high dynamic range image-based sky models in lighting simulation. Luekos, J Illumin Eng Soc (IES) 7(2):69–84

Jelle B et al (2012a) Fenestration of today and tomorrow: a State of the art review and future research opportunities. Sol Energy Mater Sol Cells 96:1–28

Jelle BP, Breivik C, Røkenes HD (2012b) Building integrated photovoltaic products: a state-of-the-art review and future research opportunities. Sol Energy Mater Sol Cells 100 (2012):69–96

Konis K (2013) Wiring to the sky. In: Adaptive architecture: proceedings of the 33rd annual conference of the association for computer aided design in architecture (ACADIA). 24–26 Oct 2013, Cambridge Ontario, Canada

Konis K, Lee ES (2015) Measured daylighting potential of a static optical louver system under real sun and sky conditions. Build Environ 92:347–359. doi:http://doi.org/10.1016/j.buildenv.2015. 04.024

LBNL (2016) Lbnl Skycam. http://flexskycam.lbl.gov/. Accessed 29 Aug 2016

Lee ES, Fernandes LL, Goudey CH, Jonsson CJ, Curcija DC, Pang X, DiBartolomeo D, Hoffmann S (2013) A pilot demonstration of electrochromic and thermochromic windows in the Denver Federal Center, Building 41, Denver, Colorado. LBNL Technical report LBNL-1005095. http://eetd.lbl.gov/publications/a-pilot-demonstration-of-electrochr-0

Lee ES, Gehbauer C, Coffey B, McNeil A, Stadler M, Marnay C (2015) Integrated control of dynamic facades and distributed energy resources for energy cost minimization in commercial buildings. Sol Energy 122(2015):1384–1397

Okalux (2016) http://www.okalux.de/okalux-gmbh-home/. Accessed 29 Aug 2016

Sage Glass, LLC (2016) http://sageglass.dreamhosters.com/technology/how-it-works/. Accessed 29 Aug 2016

SunCentral (2016). http://www.suncentralinc.com/. Accessed 29 Aug 2016

The Connected Lighting Alliance (TCLA) (2016) http://www.theconnectedlightingalliance.org/. Last accessed 23 Oct 2016

U.S. Department of Energy (DOE) (2011) Building energy data book. https://catalog.data.gov/dataset/buildings-energy-data-book

View Glass (2016). http://viewglass.com/assets/pdfs/igu-data-sheet-us.pdf. Accessed 29 Aug 2016

Ward G, Mistrick R, Lee ES, McNeil A, Jonsson J (2011) Simulating the daylight performance of complex fenestration systems using bidirectional scattering distribution functions within radiance. Leukos J Illumin Eng Soc North America, pp 241–261

Chapter 4
A Performance-Based Design and Delivery Process

4.1 Introduction

The emergence of low energy and Zero Net Energy (ZNE) building performance requirements combined with a growing array of human-factors objectives for light is driving a reversal of the conventional process of design and performance analysis. Rather than using a predetermined design as a starting point for analysis, practitioners and researchers are exploring how performance requirements can be used to identify promising solutions among multiple early-stage design alternatives. In an ideal case, exploration begins in the earliest stages of conceptual design, enabled using iterative, simulation-based analysis and informed by emerging "form-finding" workflows. In a conventional design process, energy/environmental analysis tools are rarely used to inform design decision-making in early stages of design, if at all. Rather, analysis occurs after design development, often for code-compliance purposes or to obtain green building certification. Consequently, feedback from analysis cannot be usefully incorporated into changes to the project that may improve comfort and energy efficiency. Because the largest impacts on project performance are generally established by decisions made in early stages of design, it is critical for performance evaluation to be integrated into the conceptual and schematic phases of design, where significant changes can be made without large impacts on project cost or schedule. Furthermore, decision-making about design doesn't stop at the construction documentation stage but may continue through ongoing value engineering, construction, outfitting and commissioning of the final building. It may even continue to the stage where new occupants experience the space and learn how it is designed to support their work. These activities can be facilitated by the use of various types and scales of physical mockups, beginning early in design schematics and continuing into the construction phase to fine tune the interactions between a variety of integrated systems, their controls and the building's occupants.

© Springer International Publishing Switzerland 2017
K. Konis and S. Selkowitz, *Effective Daylighting with High-Performance Facades*, Green Energy and Technology, DOI 10.1007/978-3-319-39463-3_4

Whole-building performance specifications, building energy benchmarking and disclosure requirements, outcome based codes and energy-performance-based procurement add additional incentive for design teams to seek mechanisms for reliable, performance feedback throughout all stages of design and project delivery. Climate also plays a critical role in the performance-based design process. In addition to the integration of on-site energy harvesting technologies, projects targeting low or ZNE outcomes often implement passive environmental control strategies (e.g. solar control, natural ventilation, thermally charged/discharged mass, daylighting), which must be carefully designed in response to local climate and context. Therefore, simulation tools must be capable of reliably modeling the effects of the local climate and urban context as well as the behavior of passive systems and occupant impacts. Finally, the shift towards environmentally responsive design strategies places a renewed focus on the role of building occupants in project performance, both in terms of long-term acceptance of comfort conditions in more dynamic indoor environments, as well as in terms of occupant interaction with the building energy concept.

4.2   Performance-Based Design

Performance-based design is an iterative process focused on balancing whole-building energy reduction targets with a range of Indoor Environmental Quality (IEQ) goals, while addressing classic programmatic constraints of time and cost. There is no single optimization tool available to translate project objectives and constraints into a single design outcome. Nor is there consensus for how to best manage trade-offs among various performance objectives (as discussed in Chap. 2), or how to assign relative weighting to indicators based on their perceived importance among various project stakeholders (e.g. design team, project manager, or end users). In the real world, designers must sort the global problem into chunks that can be analyzed and optimized using available tools and guidance and then recombine those chunks into a coherent overall package. This is an ongoing process- it needs to be initiated at one level of detail in early design/schematics and then continued later through design development, construction documents, value engineering and even late in construction.

As noted at the beginning of the chapter, energy/environmental analysis tools are rarely used to inform design decision-making in early stages of design, if at all. For example, even within the limited group of architectural design firms committed to meeting the ambitious goals of the Architecture 2030 Commitment (the American Institute of Architects' (AIA) program to quantify and report the progress AIA members are making in the effort to reduce greenhouse gas emissions in the built environment), only 44% of projects used energy modeling at all in concept/schematic design. And only slightly more than half (55%) used energy modelling at the later stages of design (design development and construction documents) (AIA 2014). When energy modeling does occur it is frequently too late in the process to

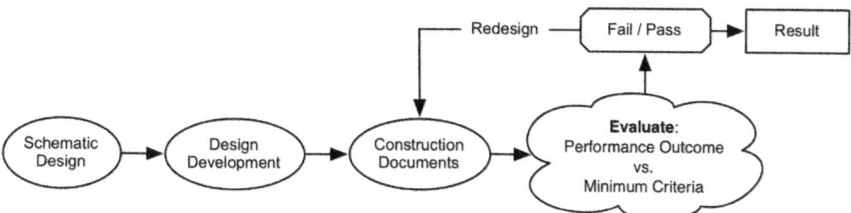

Fig. 4.1 A compliance-based design process

have maximal impact, e.g. after design development, often for code-compliance purposes or to obtain green building certification (Fig. 4.1). Responsibility for modeling and analysis is often shifted outside the discipline of architecture to lighting designers or to Heating, Ventilation and Air Conditioning (HVAC) engineering consultants. Given the fragmented nature of design, their focus has traditionally been on meeting minimal standards with incremental improvements in building assemblies and systems efficiency guided by energy standards (e.g. American Society of Heating and Refrigeration Engineers ANSI/ASHRAE/IES Standard 90.1-2013). Similarly, due to the complexity and resources required to conduct photometrically-accurate lighting simulations, lighting design is often performed by external consultants and late in the design process, where the emphasis is placed on specifying high-efficiency electrical lighting fixtures and photo-sensitive lighting controls for a largely-completed daylighting design concept.

This slow and fragmented approach limits the potential to explore architectural strategies (e.g. siting, building geometry, window-wall ratio, shading devices, etc.) to minimize heating and cooling loads and the application of environmental services such as natural ventilation, exposed thermal mass, and daylighting as passive alternatives to conventional HVAC and electrical lighting systems. To effectively meet low and ZNE performance objectives, workflows are needed that enable designers to examine and optimize the application of passive environmental strategies and add and optimize the daylighting controls needed to optimize performance. And feedback is most valuable if it can be generated in "near-real-time" to keep up with the pace of decision-making both in early-stage design and all the way through the value engineering stage. The emergence of simplified user-interfaces and "plug-ins" linking complex physically accurate energy and lighting simulation tools with the 3D authoring software used by designers has helped to facilitate the adoption of performance-based design by architects, while raising a new set of issues related to the reliability, visualization, and appropriate interpretation of simulation-based outcomes.

The building industry finds itself in transition in terms of responsibility, tools and functions. Historically each member of the design team had a well-defined niche in which they worked, with a set of narrow specialized, compartmentalized tools. Handoffs between team members at design reviews would be accompanied by related translations of software outputs from one set of specialized design or analysis tools to another. Needless to say this was slow, costly and error prone so analysis was minimized. Over the last two decades paper based drawing and design has transitioned to Computer Aided Design (CAD), and more recently the underlying geometry models shared between architects have now been extended from 2D to 3D and are used by many or all members of the design team with their domain expertise added to the data model. The original geometric building model has now grown to become a more comprehensive Building Information Model (BIM), which can be used across design specialties and throughout the design process, during construction and even for operations by facility managers. Issues of standardization and interoperability are still being worked out, but global industry interest in this evolving model is high and is likely to continue in the years ahead.

In contrast to the conventional process shown in Fig. 4.1, a performance-based design process is defined by a feedback mechanism utilizing analysis tools to relate design decisions to explicit, measureable project performance outcomes (Fig. 4.2). By examining how design decisions impact project performance, knowledge can be generated and fed back to inform decision-making to improve the performance of future design iterations, and ultimately to arrive at the final design goals.

One obvious fact that is often overlooked is that performance-based design is largely dependent on (1) how well "performance" goals are defined by the design team in terms of specific, assessable performance indicators, (2) the existence/capabilities of analysis tools to accurately calculate the chosen performance indicators, and most importantly (3) the *validity* of the performance criteria used to define "effective" performance. The term "validity," in this context, refers to

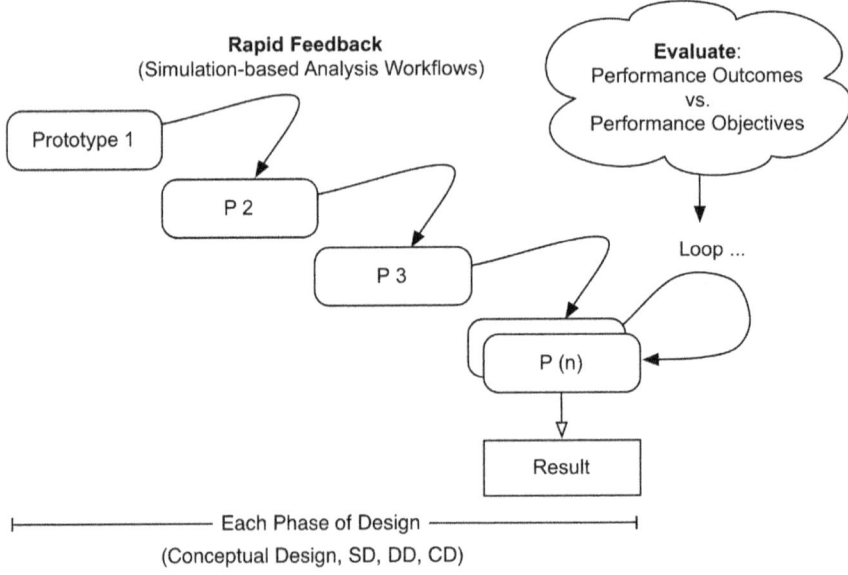

Fig. 4.2 A performance-based design process

the appropriateness, aggressiveness, and comprehensiveness of performance criteria. Without metrics, multiple design strategies cannot be compared and improvement cannot be determined. Therefore, indicators must be established for each objective as well as criteria for interpreting what constitutes "effective" or "optimal" performance. Indicators for energy performance (e.g. annual Energy Use Intensity (EUI), peak demand) are well defined and their use is already established in practice. There is less consensus for what performance indicators are appropriate for evaluating IEQ factors related to daylighting, and what criteria must be met to ensure satisfied and comfortable occupants as well as less wide-spread adoption in practice. A review of performance indicators is provided in Chap. 2.

A performance-based design process is most useful if it leads to designed outcomes that perform as anticipated. Consequently, the performance criteria established during design must be assessable post-occupancy, to ensure that performance in use can be evaluated against design intent. In addition, the assumptions for occupant comfort and satisfaction underlying various performance indicators and objectives require validation to ensure that meeting these objective criteria leads to comfortable and satisfied building occupants. This latter task is addressed in Chap. 6.

4.3 Simulation-Based Design Tools and Workflows

With the growing complexity of design solutions and the need to predict performance hourly throughout the year and to address occupant comfort as well as energy, the role of design tools becomes more important, as does the skill of the users and the quality of the input data that drives them. "All tools are wrong, but some are useful" is a valuable starting guide to the use of simulation tools in the design process. Design tools, in this context, refer to individual software programs that provide specialized feedback on specific aspects of performance (e.g. window thermal and optical performance, two-dimensional heat transfer, visual comfort, or annual heating/cooling energy use). These tools are only the first step in a long sequence of decision-making leading to performance outcomes in the built, occupied building. Translating a design goal into measured performance requires many steps, each with their own inputs, uncertainties and outputs, as noted in Fig. 4.3. The effective use of simulation tools is an essential precursor to achieving good outcomes but only one of many factors that ultimately impacts performance.

The need to examine trade-offs between various energy and IEQ factors, as well as understand the whole-building performance of various integrated systems and control options is driving software developers and designers to integrate 3D authoring software, highly optimized energy and lighting simulation engines (e.g. EnergyPlus and Radiance), and parametric modeling and optimization tools into generative design *workflows*, utilizing visual scripting to rapidly adjust one or more building parameters and visually and quantitatively understand performance feedback in near-real time. A workflow, in this context, refers to a general step-wise

Fig. 4.3 Measured
performance versus design
goals

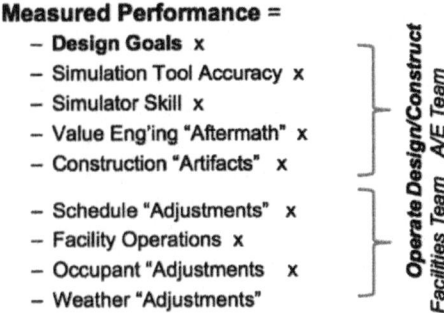

process of setting quantitative performance objectives, creating a BIM model consisting of building geometry, climate and site information, occupancy and operational schedules, material properties and building environmental systems, then simulating the annual performance of the model to calculate one or more performance indicators. Outcomes are then compared to design intent, through a variety of numerical and visual displays of information and the building model is revised based on confidence in available feedback (for an early example, see Andersen et al. 2008[1]). This is increasingly becoming an iterative process, where several iterations of a model are needed in a short time to effectively inform early stage design decisions. A workflow necessitates the use of multiple software modeling and simulation tools, as well as additional software to post-process, visualize and interpret results. Where iterations are performed, additional optimization software or purpose-built scripts may be applied to automate the execution of a large number of iterations seeking pre-defined goals. These can be executed using at least three different approaches: (1) trial and error variations in design/input parameters based on the experience of the design team; (2) massive "brute force" parametric simulation of key variables across the design parameter space searching for solutions that meet key performance objectives; (3) genetic algorithms or structured optimization methods that seek minima or optima in performance outcomes using specialized optimization toolkits.

4.3.1   Life-Cycle Building Information Model

In an ideal *design-deliver-commission-disclose* scenario, a BIM model is created and maintained across the design process and is available to, and compatible with, all the domain models being used by all team members. Models are later used for cost estimation, code compliance and building rating systems. But that should be just the start as the ideal situation would also make the same models available to the contractor if change orders were discussed and to the commissioning team as the final

---

[1]An intuitive daylighting performance analysis and optimization approach.

operational touches are put in place. The long-term vision is a life cycle data model in which the operator inherits a corrected, as-built model of the building to assist in operations over time and to continuously determine if performance expectations are met during operations. Lessons learned from the project in use, disclosed and shared systematically, are then accessible to inform future design projects, human-factors models, and technology specifications (among a range of other potential uses).

4.3.2   Contextual Awareness

Predicting solar radiation and daylight exposure over the course of a year is a critical first step in understanding solar control requirements and daylight availability for the building facade, with implications for the orientation and footprint of the building, the organization of interior program, glazing selection, the location and sizing of window apertures, the configuration of exterior shading systems, and a number of other considerations. Analysis is particularly important in the early stages of design for projects that seek to implement passive or low-energy indoor climate strategies, which are limited by lower peak thermal loads compared with conventional HVAC strategies. At the building scale, radiation mapping is a useful tool to understand the critical areas of the building envelope for solar control, as well as the daily and seasonal variation in available solar energy, which may inform the location and sizing of solar thermal or solar photovoltaic systems. The approach is particularly applicable in dense urban environments to inform planning for outdoor sun exposure and thermal comfort, urban heat island mitigation, and to assess the potential for negative impacts of the project on existing structures. We note that while direct radiation from the sun is important, sky diffuse solar radiation and ground or building-reflected radiation may also be critical to some designs. We also distinguish between solar radiation modeling and daylight modeling, which while related by atmospheric effects are often assessed by different tools and processes.

Significant errors can be introduced if the manner in which the urban context mediates between the project and the sky vault is not appropriately considered in the design process. To address these issues, a number of efforts have been made in recent years (Compagnon and Raydan 2000; Mardaljevic and Rylatt 2000; Robinson and Stone 2004; Ashdown and Ward 2013) to improve the ability to map irradiation while taking into consideration the complexity introduced by local weather conditions, adjacent obstructions in the urban context and the practical need for rapid annualized simulation results for multiple schemes. The increasing availability and organization of contextual information, such as standardized climate data files,[2] publically available Geographic Information System (GIS) data, and complete digital 3D models of urban building geometry is enabling designers to develop workflows where project design can be directly informed by local site and climatic conditions.

---

[2]http://mostapharoudsari.github.io/epwmap/.

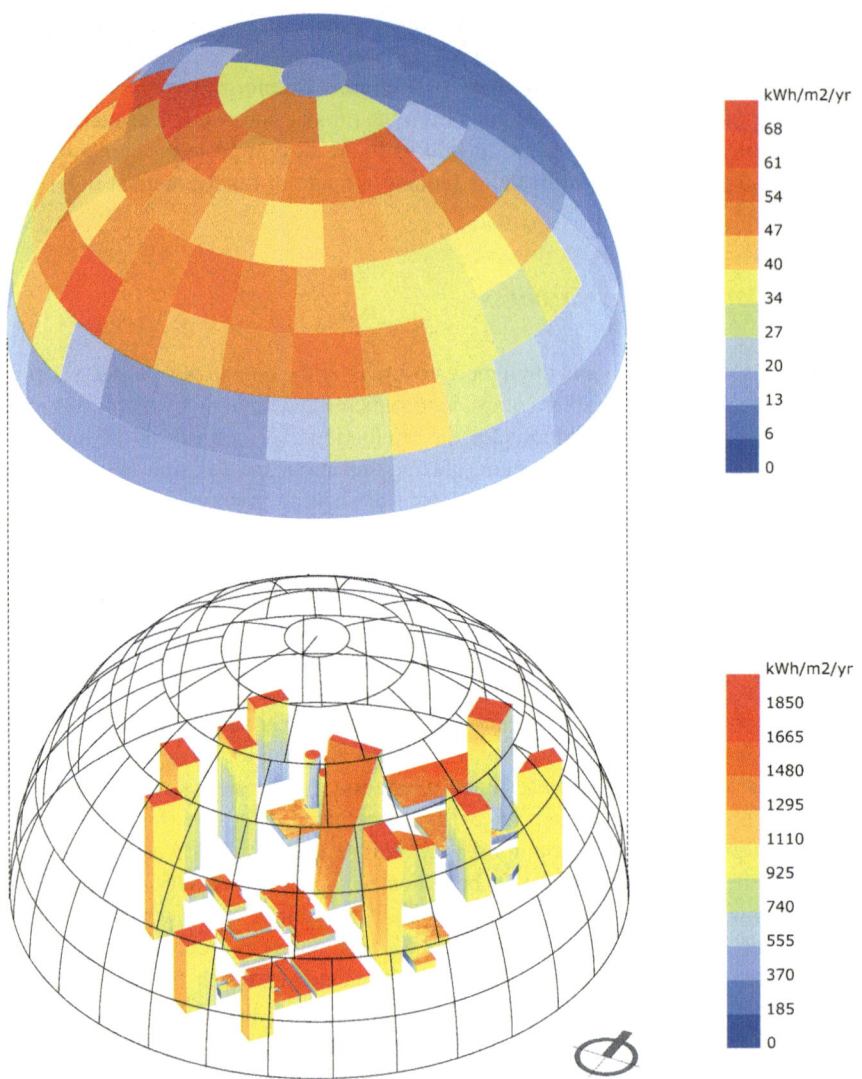

Fig. 4.4 Hemispherical Tragenza sky matrix with 145 patches (*above*) used to calculate total annual solar irradiation (kWh/m²/year) on urban building geometry in downtown Los Angeles (*below*)

The example presented in Fig. 4.4, implemented using the Ladybug skydome and radiation analysis components (Roudsari and Pak 2013), follows the technique develop by Robinson and Stone (2004) for simulating solar irradiation on the urban fabric. The Robinson and Stone technique, called *GenCumulativeSky*, discretizes the sky vault into a set of patches, which subtend a similar solid angle. The pattern shown in Fig. 4.4 is a Tragenza sky matrix, developed by Tregenza and Sharples (1993) where the sky vault is divided into seven azimuthal strips and where each

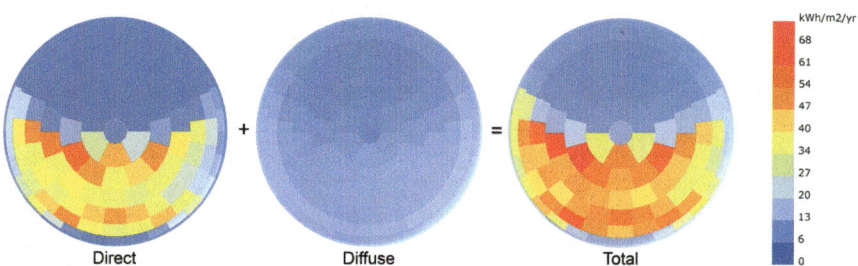

Fig. 4.5 Top view of hemispherical Tragenza sky matrices for Los Angeles showing distribution and magnitude of unobstructed direct, diffuse, and total radiation

strip is subdivided into patches each with a similar solid angle, producing 145 total patches. Using solar data from a geographically appropriate Energy Plus Weather (EPW) file, the radiance at the centroid of each patch is calculated using the Perez all weather luminance distribution model (Perez et al. 1993) and the results for each hour of the analysis period (e.g. a series of days, months, or a full year) are aggregated into 145 patch bins. Given the set of sky patches, patch solid angle, patch radiance, the mean angle of incidence between the patch and the analysis plane (modified by the proportion of the patch that is in view), the annual contribution of diffuse and direct irradiance can be calculated and visualized (Fig. 4.5). However, the technique does not account for the contribution of indirect (e.g. reflected sunlight), and becomes problematic for analysis of highly reflective surface geometry. The example in Fig. 4.5 was produced using *gendaymtx*, a Radiance program similar to *GenCumulativeSky* which, in addition to the Tragenza matrix, implements Reinhart's extension of the Tregenza sky, where the original 145 patches can be subdivided into a user-defined number of sub-patches (e.g. 16), except at the zenith, to enable significantly greater resolution of the sky vault, and more accurate sun penetration data which is particularly useful in those design cases where this level of detail may be important, e.g. museum glazing design.

4.3.3   Building Form and Form-Finding Workflows

Traditional analysis-supported design starts with a design concept and the analysis is used to understand the performance and informs potential improvements to the design. The emerging availability of new tool sets and workflows offers a chance to invert the process and use the tools to create designs. New open-source environmental analysis plug-ins, which link advanced thermal and lighting simulation engines with parametric 3D authoring tools and genetic optimization tools has enabled the development of novel "generative" (e.g. Caldas 2008; Gagne et al. 2012) or "form-finding" workflows. Rather than analyze and attempt to improve the performance of a predetermined design, these workflows require the designer to pre-define multiple performance objectives and then offer the potential to explore a

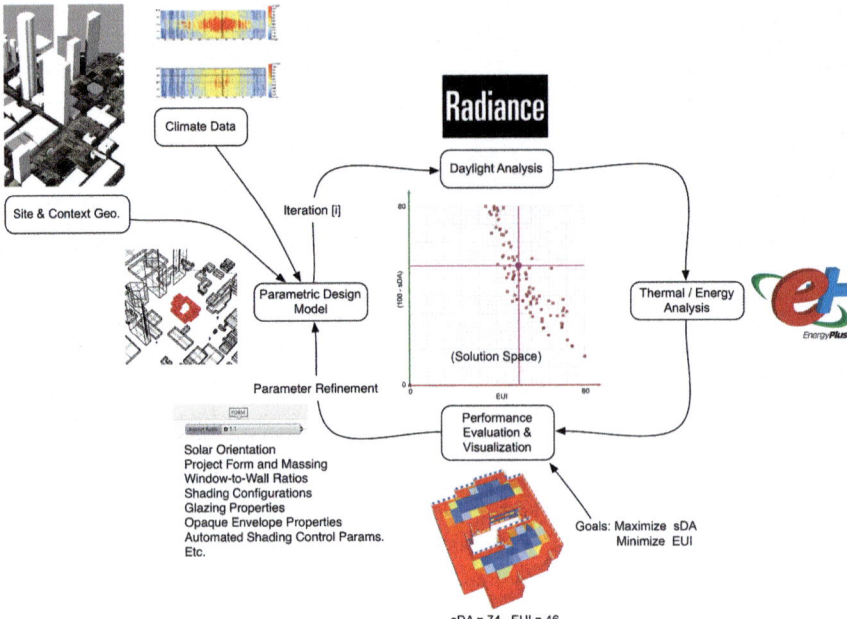

Fig. 4.6 Workflow diagram showing primary components and iterative analysis and evaluation feedback loop

solution space of possible design solutions generated automatically using genetic algorithms. The following sections outline one approach to the application of form-finding workflows to inform early-stage decision-making.

The workflow presented in the following sections was developed using *Grasshopper*,[3] a visual algorithm editor that runs within the Rhinoceros 3D CAD application. The goal of the workflow is to provide rapid feedback on the daylighting potential of various building forms generated from a variety of input parameters commonly explored in early stage design (e.g. building footprint, solar orientation, massing, number of stories, building fabric and fenestration). A detailed description of an earlier version of the workflow is provided in (Konis et al. 2016). The workflow is organized into steps as shown in Fig. 4.6, which begin with basic data input related to project climate and site information. The user then defines geometry for a preliminary building model by adjusting a predefined set of design parameters. Once a model is generated, annual climate-based daylighting and thermal energy simulations are run and quantitative performance outcomes are produced and displayed visually. Results can then compared to outcomes from a reference building situated on the same site, such as an ASHRAE 90.1 compliant reference building from the U.S. Department of Energy (DOE) set of commercial building reference models (Deru et al. 2011) or another appropriate reference.

---

[3]http://www.grasshopper3d.com/.

Because the building form is generated parametrically, model parameters can be explored in an ad hoc fashion and simulations re-run to compare performance outcomes for various building form and fenestration combinations. This use of the workflow may be sufficient for preliminary exploration of a handful of design alternatives. However, even with a small number of parameters, evaluation of all possible parametric combinations can quickly lead to the need to perform millions of annual simulations. This challenge is addressed through the introduction of an optimization component, which applies evolutionary multi-objective optimization principles to automatically provide a range of optimized trade-off solutions between the extremes of each performance objective defined by the designer.

4.3.3.1 Site

As discussed earlier, outcomes from both energy and daylighting simulations are influenced, often significantly, by the presence of surrounding building geometry. For early stage design, it is critical to examine design prototypes that are appropriately situated (and modeled) in the built context of the project site. To facilitate the rapid modeling and visualization of the urban context, the workflow integrates the plug-in *CADtoEarth* (AMC Bridge 2016) to automatically import Google Earth images and building geometry from an open-source GIS repository[4] into the 3D authoring workspace (Fig. 4.7). Surrounding buildings are automatically represented as shading objects in both the thermal energy analysis (EnergyPlus simulations) and the daylighting analysis (Radiance simulations). Note that these approaches may

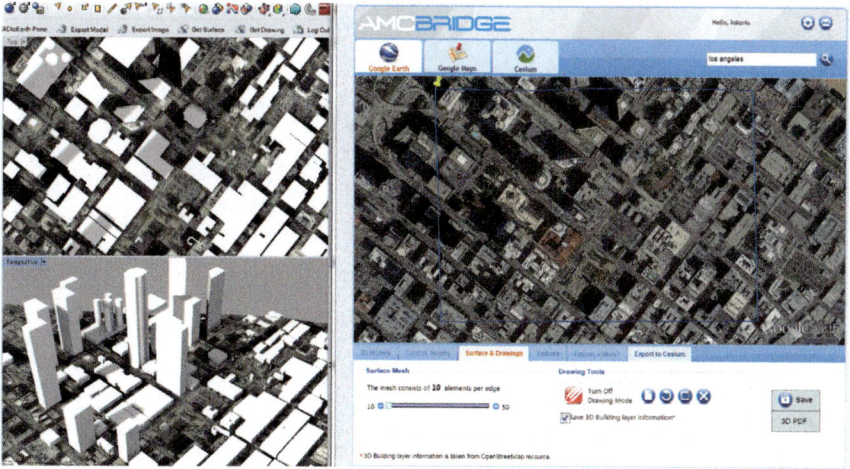

Fig. 4.7 Example use of *CADtoEarth* to import existing site information for a hypothetical project site in downtown Los Angeles into 3D authoring environment

---

[4]www.openstreetmap.org.

import the 3D geometry but do not directly address other modeling parameters, such as surface reflectivity.

4.3.3.2 Parametric Building Model

The basic geometry of a project prototype can be generated using two different approaches. The first approach is to define one or more building footprints (as polylines) along with a total project floor area. The workflow will then automatically generate a building volume that accommodates the specified floor area using a default assumption for the number of floors. The total number of floors can also be explored parametrically, where the workflow will automatically adjust the project footprint to accommodate the specified number of floors. Thus, each original footprint becomes the basis for a wide range of formal variations, each of which can then become the basis for parametric variation of project orientation, WWR assigned to each facade, the application of various facade shading strategies, and many other parameters. The first approach assumes that the designer has some initial concept for the building footprint on the project site. The second approach, presented here, removes this assumption and operates by applying a "form-finding" component to automatically generate more complex shape boundaries including courtyards while automatically adjusting the building volume to maintain the specified project floor area. These parameters (Fig. 4.8) are automatically adjusted by an evolutionary

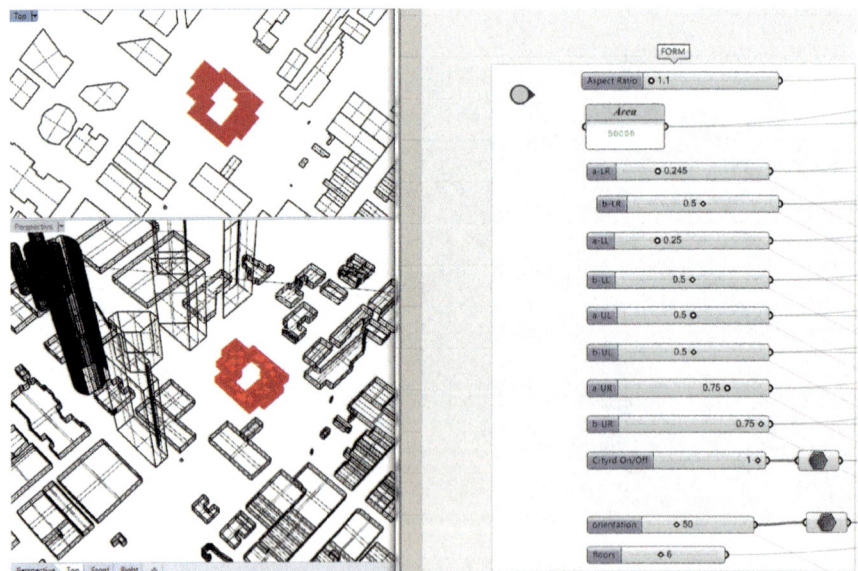

Fig. 4.8 Grasshopper sliders used to adjust building parameters such as number of floors, solar orientation, courtyard size and shape, etc. for a fixed building area (screen capture from Grasshopper)

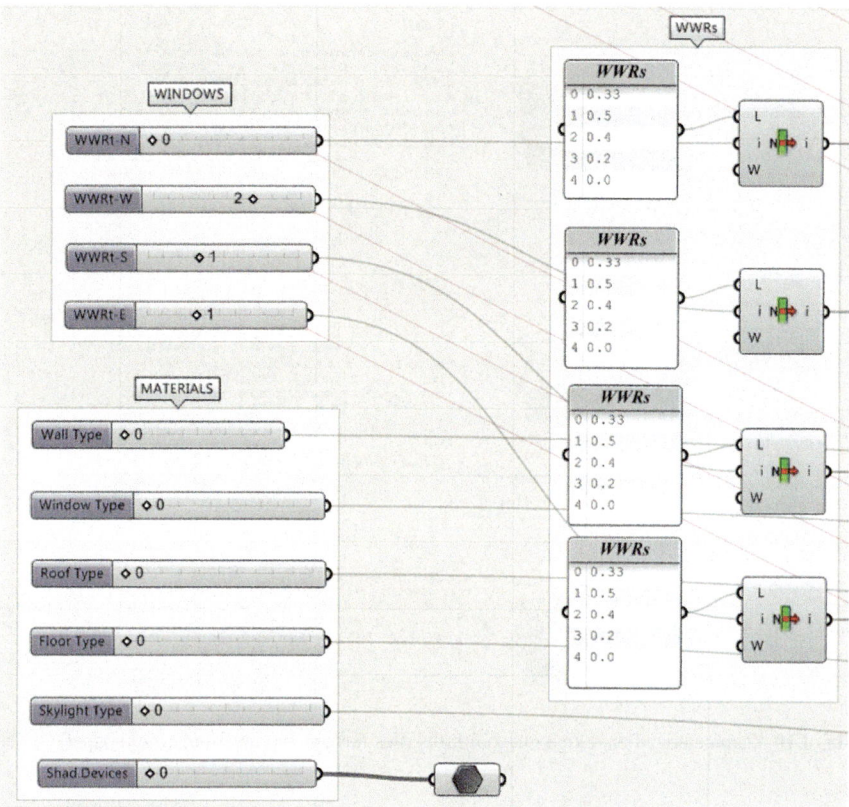

Fig. 4.9 Parameterized window-to-wall ratios for each facade orientation (screen capture from Grasshopper)

multi-objective optimization component *Octopus* (Vier 2016) to enable the generation of forms with the objective of maximizing daylighting and minimizing EUI within the space of possible solutions generated by the form-finding component. A unique Window-to-Wall Ratio (WWR) parameter is assigned to each facade of a given formal prototype (Fig. 4.9), which can be explored in combination with various fenestration and glazing systems (Fig. 4.10) to enable the optimal WWR for each facade to be determined for a specified climate and site condition.

Exterior solar shading elements can be automatically generated in response to solar vectors determined by specifying the range of hours during the year that direct solar control is required. This deterministic approach differs from the form-finding approach used to explore and optimize other building parameters such as form and solar orientation. While a more sophisticated approach may consider embedding form-finding of individual shading elements within form-finding of the overall building shape, this was viewed as a stage of refinement that is better applied as a second phase (or scale) of optimization after preliminary building forms have been identified and prioritized. Figure 4.11 shows the exterior shading elements

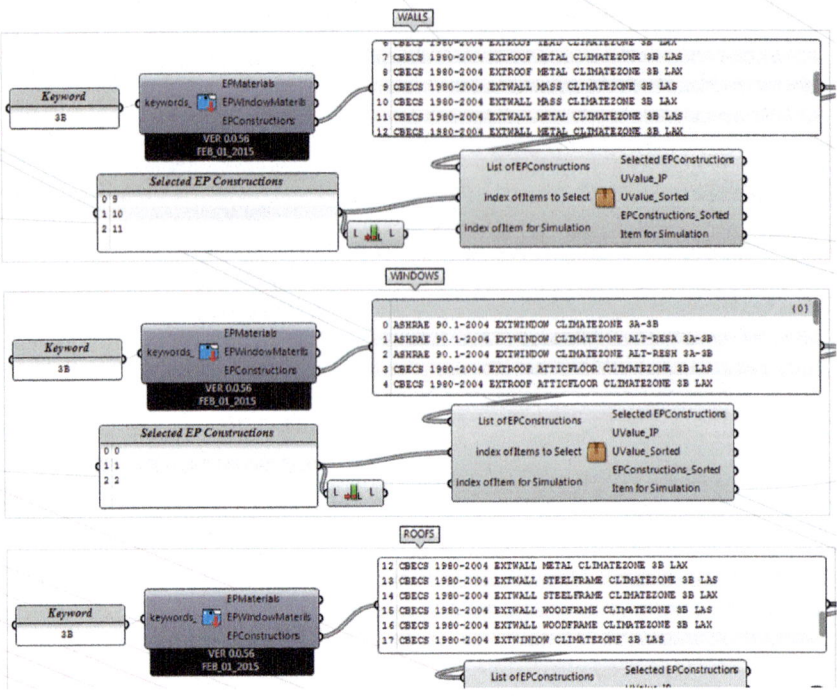

Fig. 4.10 Parameterized envelope material properties (screen capture from Grasshopper)

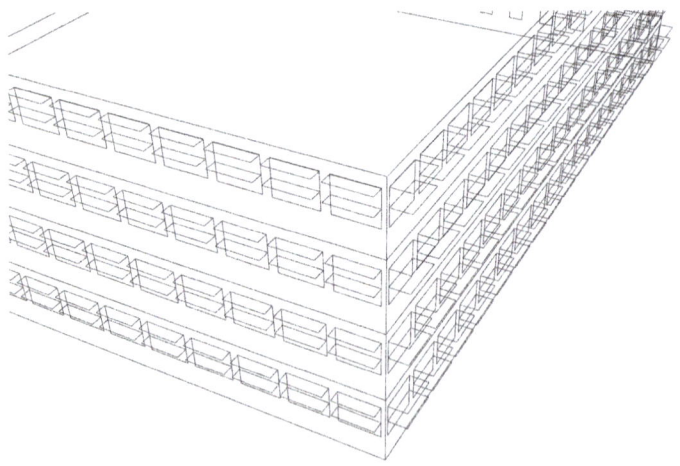

Fig. 4.11 View of exterior shading elements automatically generated for windows on the south and east facades of a project located in Los Angeles, CA

automatically generated (based on the deterministic approach outlined above) for windows on the south and east facades of a project located in Los Angeles.

4.3.3.3   Daylighting Analysis Using Radiance

Daylighting analysis is executed utilizing the plug-in Honeybee, which interfaces with the lighting simulation engine *Radiance* (Ward and Shakespeare 1998) and the Radiance-based program Daysim (Reinhart 2016) to enable annual hourly climate-based simulations. It is important to note that EnergyPlus includes its own hourly, climate-based simulation engine for daylight, which is suitable for shallow spaces and simple glazing materials. The workflow utilizes the metric Spatial Daylight Autonomy (sDA) to quantify and evaluate annual daylighting performance. As discussed in Chap. 2, sDA describes annual sufficiency of ambient daylight levels in interior environments (IES 2012). A grid of analysis points located 0.7 m above the floor is used to assess sDA_{300}. An example outcome is presented in Fig. 4.12. The false-color mapping indicates the level of

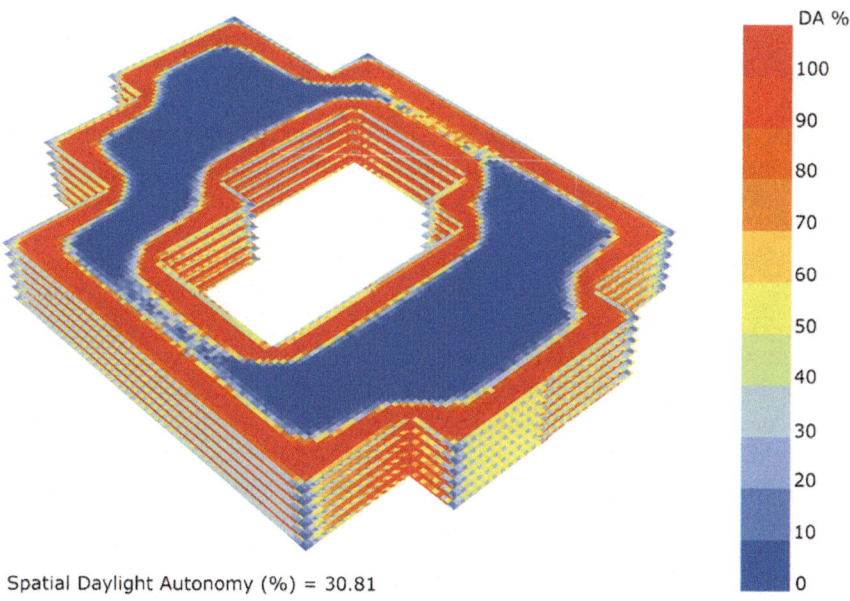

Spatial Daylight Autonomy (%) = 30.81

Fig. 4.12 Visualization of annualized simulation outcome for a 6-floor, 50,000 m^2 commercial building using false-color scale to illustrate Daylight Autonomy (DA) outcomes for each analysis point with a 300 lx threshold

Daylight Autonomy (DA) achieved for a particular location, ranging from 0 to 100% on an annual basis, where blue indicates poor performance (few or no hours of DA) and red indicates good performance, e.g. many (or all) hours achieve DA over an annual period. A count of all DA scores that exceed a threshold of 50% is taken to compute the final sDA outcome for the building of 30.8%. In this example, 30.8% of all locations achieve a DA of 50% or more of occupied hours during the year.

4.3.3.4   Thermal/Energy Analysis Using EnergyPlus

The plug-in Honeybee (Roudsari and Pak 2013) is also used to run annual building energy simulations by supplementing the building model with information describing the various envelop material properties, occupancy and equipment schedules, operational assumptions, climate and site conditions required for detailed EnergyPlus simulations. Following each annual daylighting simulation using Radiance and Daysim, an hourly annual simulation was performed by EnergyPlus to quantify the annual Energy Use Intensity (EUI) incorporating all energy use impacts including the potential benefits of natural ventilation. Prior to the EnergyPlus simulation, daylighting results were interpreted to generate custom electrical lighting schedules for each zone which are then passed to EnergyPlus to account for the electrical lighting energy reductions achieved by effective daylighting. The natural ventilation approach implemented is a "mixed-mode" approach, which consists of using design geometry and climate data to estimate the periods during the year where natural ventilation could meet the cooling demands of the indoor space, thereby reducing the annual operating hours of mechanical HVAC.

4.3.3.5   Visualization

The outcomes of individual design iterations can be visualized in real time as they are produced during the optimization process and can be visually aggregated based on performance outcomes to gain insight into the formal characteristics that lead to both performance improvements and decrements with respect to the "current" design solution. Figure 4.13 shows the variation in building form, number of floors, orientation and window configurations being explored for the first 140 iterations of a 50,000 m^2 commercial office building located in Los Angeles, CA, situated within the urban context shown in Fig. 4.7. The same falsecolor scale used in Fig. 4.12 is used in Fig. 4.13 to report DA (ranging from blue = 0% to red = 100%). The outcomes are ordered (top to bottom, left to right) from the best combined

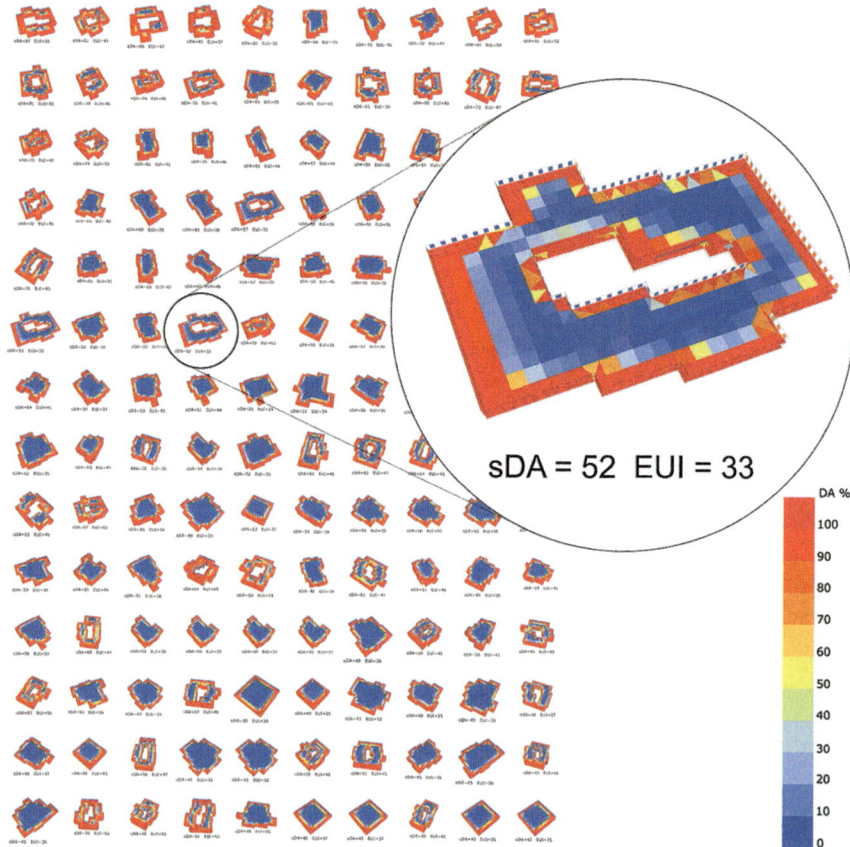

Fig. 4.13 The first 140 simulation iterations of a form-finding optimization for a 50,000 m² commercial office building located in Los Angeles, CA

performance (greatest sDA and lowest EUI) (upper left) to the worst combined performance (lower right).

4.3.3.6 Optimization

Evolutionary multi-objective optimization is implemented using the Grasshopper plugin *Octopus* to explore various combinations of parameters and examine outcomes relative to one or more performance goals. The best trade-offs between the specified objectives are searched, producing a set of possible solutions.

Figure 4.14 presents the two-dimensional solution space (sDA vs. EUI) after two generations of solutions have been completed. The performance of the "base case" model is highlighted with a purple circle. Four quadrants can be defined (purple

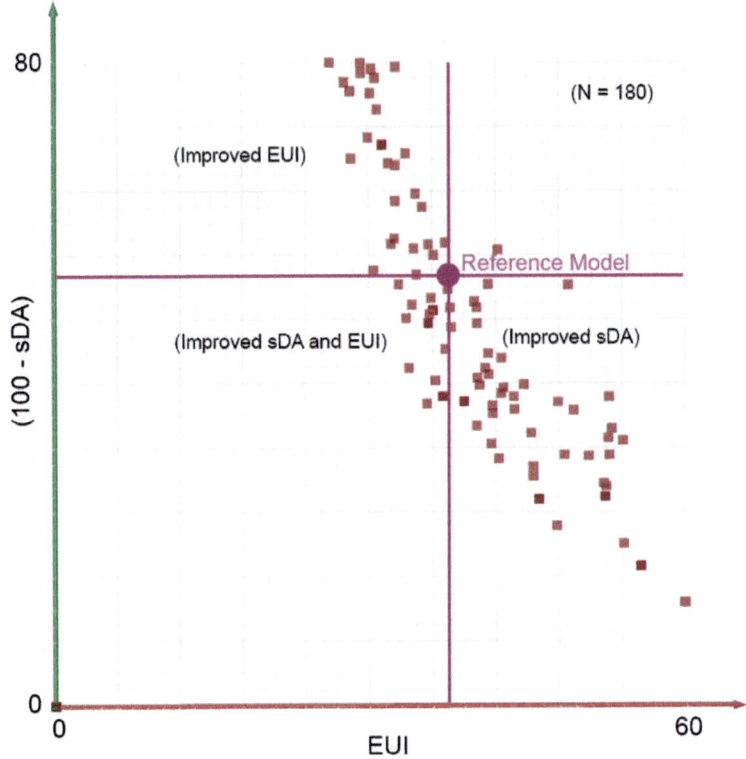

Fig. 4.14 Two-dimensional solution space showing performance results for two generations (100 iterations/generation) for a 50,000 m^2 commercial office building situated in a downtown Los Angeles site

lines) to identify solutions that result in either improved or reduced performance relative to the basecase in terms of daylighting (sDA) and EUI. Outcomes within the solution space can be easily identified and reinitiated in the 3D authoring environment for visualization and further refinement.

4.3.3.7   Discussion

The solutions generated by the workflow can not only help to refine final design solutions, but also serve to define more aggressive reference targets for design development, where further refinement could be achieved by more focused evaluation of specific attributes, such as optimized exterior solar shading. These studies reaffirm past experience that significant improvements in daylight availability achieved by relatively simple adjustments to building orientation and geometry and in fenestration configurations, which highlights the importance of exploring building form in early stage design. A validation study documented in Konis et al.

(2016) demonstrates that the daylighting and energy form-finding workflow is capable of generating design solutions that are significantly better than an ASHRAE 90.1-compliant reference building across a range of climates and urban settings.

The performance improvements demonstrated in the present example where achieved with relatively few values (between 2 and 4) set for a limited number of building parameters. Even with this preliminary level of parametric variation, over 8 million design solutions exist, making it impossible for the user to explore the solutions space without an automated workflow. More refined outcomes could be achieved by including a greater number of parameters and intermediate values along with the simulation of additional generations of solutions at the expense of increased simulation time and/or the need for more computational resources. However, it is important to note that achieving successful optimization outcomes requires the designer to specify appropriate design parameters and constraints to avoid simulation of potentially millions of design outcomes that would never be realistically considered by the project team. In real projects, many key parameters are constrained by the project site, program, or preferences of the client. However, in this "real world" context, optimization workflows have the potential to enable the design team to explore the range of solutions available within those constraints. Thus, automated design has two particular values, (1) to define the envelope of possible solutions and, (2) to help refine impact of design details. Finally, the diversity of "good" solutions means that there is not just a single possible design solution that works, or performs optimally.

4.3.4 Fenestration

As outlined in Chap. 3, the building facade and perimeter zone represent a complex design integration challenge. Fenestration systems, comprising the window or curtain wall elements that hold the glazing, and any interior or exterior "attachments" represent a similar challenge. In modern buildings, glazings are no longer a simple single layer glass: they can consist of many different combinations of glazing layers, ceramic frits, coatings, gas layers, spacers, and dividers. While each glazing has specific optical properties, new "switchable" glazings have optical properties that change dynamically in response to voltage, light or heat. The glazing element (or IGU, insulated glazing unit) is then held in place in the opaque envelope with the window sash and frame, or by the structural elements of the curtain wall. Larger fenestration systems are often functionally subdivided into multiple glazing or window elements, each of which performs one or more specific roles, leading to more diverse and complex specification of components. Further, fenestration glazing and framing is normally supplemented with exterior and/or interior window attachments, such as fixed louvers or interior roller shades. These attachments add increasing optical complexity and often have the potential to be controlled dynamically either via automation or by building occupants. Finally, the building skin encloses multiple space and program types, and must serve the unique programmatic needs of each space while presenting a thoughtfully orchestrated outward appearance to address additional architectural and urban design goals.

There is a direct, but complex, connection between the thermal and visual comfort of occupants in a perimeter space and the details of the facade design. While the relationship between these facade design details and building energy consumption is often very important, it is also "secondary" and more convoluted. There are three challenges for designers to properly understand and address energy impacts of facades.

The first is the lack of a direct 1:1 relationship between the facade and energy use. While it is well-understood that the building envelope has an impact on energetic flows, when the energy consumption of a building is outlined it is normally broken into categories such as lighting, HVAC and plug loads, all end uses that directly consume energy and are thus readily measurable, but the envelope is omitted. The systems that provide light, heat and ventilation are often characterized by simple performance parameters with metrics that allow one to assess relative efficiency such as lumens/watt for a light source or a Coefficient Of Performance (COP) for a chiller. Assessing the performance of a glazed facade is a rather different matter since the system itself not only does not directly consume energy but the impact of its properties (U, Tv, SHGC) on those building HVAC and lighting systems is highly variable, depending on orientation, location, time of year, internal operations, etc.

The second challenge is the relationship of the perimeter zone that is most influenced by the facade to the overall building floor plate. It is common for an engineer to argue that facade design decisions are not important because they have only a 5–10% impact on building energy use. This situation arises in the design of conventional buildings with very large floor plates where the perimeter zone, defined as the space that is within 15–20 ft of the exterior, represents only 20–30% of the total floor area and does not include high energy use spaces such as server rooms. In this perspective very large perimeter impacts have only a small total building energy impact in relative terms. Finally, the design of projects in urban areas, partially those of high density, requires understanding the potential thermal and visual effects of the adjacent buildings on the facade as well as the impact of the new design on its neighbors. The bad news is that all this may make the design and assessment problem more complex and challenging. The potentially good news is that thoughtfully designed facades, using the methods and tools outlined above, have the potential to dramatically reduce costs and improve energy/carbon impacts, as well as providing other valuable "services" to occupants in terms of comfort and amenity.

Glass has evolved enormously over the last 40 years from a time when the primary options were tints and some highly reflective coatings. Glass comes in a wide range of thicknesses, and with optical properties that can range from the crystal clarity of low iron glass to heavily tinted and colored glass, to glazings with applied coatings that can manipulate the solar and long wave infrared spectra to tune optical properties. These basic glazing layers (see Fig. 4.15) can be assembled into multilayer laminates with new combined properties and then further integrated

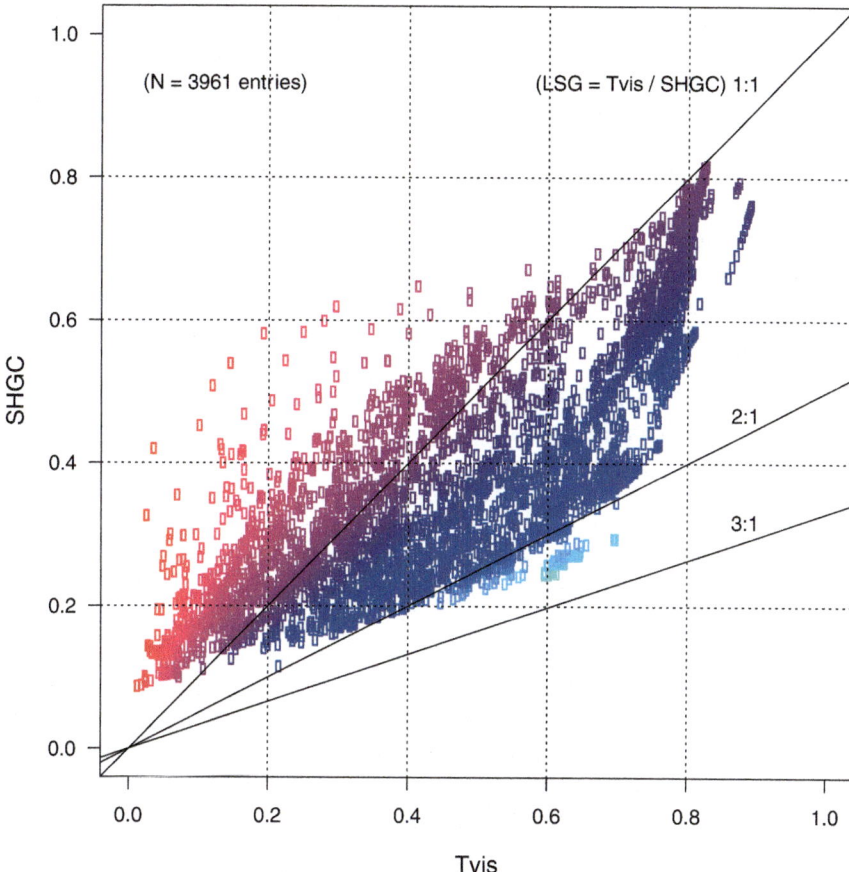

Fig. 4.15 There are over 4700 specular glazing materials in the current IGDB (v46). This chart shows the range of options for two key optical properties, visible light transmission (Tvis) and Solar Heat Gain Coefficient (SHGC) using properties for 2-layer IGUs with Argon gas fill

into an insulating glass unit with gas fills to further enhance performance properties. The resultant multitude of design options then becomes a blessing and curse for the designer. While vendors of individual products typically provide glazing or window properties (e.g. U-factors or R-values, Visible transmittance, Solar Heat Gain Coefficients or Shading Coefficients, and air leakage rates), the relative importance of these properties depends on site and building-specific conditions as well as occupant needs.

In terms of energy, considerations such as peak cooling load, peak electricity demand (and potential for load shed in a demand response event), and annual energy consumption are all influenced by fenestration design choices. However, performance outcomes for a given design vary, often considerably, depending on factors such as local climate, the unique solar orientation and urban overshadowing conditions of individual windows, and how operable elements of the system are automated or controlled manually by occupants. Adding to the challenge, designers must manage trade-offs between energy performance objectives with human factors objectives such as visual comfort, daylight availability, visual connection to the outdoors, and personal control. As discussed throughout this book, the complexity of the challenge often leads designers to apply "rules of thumb," precedent from past projects or case studies, or simplified calculations or simulation tools that discount important factors such as photometrically accurate transmission of light through optically complex fenestration layers, human-factors performance indicators such as visual comfort, or models of occupant shade control behavior.

4.3.4.1   Tools, Material Libraries, Virtual Components and Rapid Performance Feedback

In early stage design, analysis tools such as COMFEN (LBNL 2016a, b, c), can be used to guide key fenestration design decisions such as the selection of optimal glazing properties, window-to-wall ratio, evaluation of exterior shading systems, controls for automated shading, and glare control strategies. COMFEN, (which stands for Commercial Fenestration), is an interface that allows access to most facade oriented design features in EnergyPlus. COMFEN supports the design team in rapidly calculating the energy demand, daylighting impacts and comfort impacts of window/facade options in commercial buildings. A simulation-based tool such as COMFEN can be used early in the design process to help designers identify the most effective fenestration system for the unique performance goals of their project. COMFEN allows designers to quickly create various fenestration and room design strategies and then rapidly evaluate the effects of key fenestration variables including glazing, framing and shading, on energy consumption, peak energy demand, and thermal and visual comfort over an annual period (Fig. 4.16) for a wide range of climates, as well as allowing the user to drill down to look at hourly and monthly performance profiles. Simulations are performed using validated dynamic energy and lighting simulation engines (EnergyPlus and Radiance) and results are presented in graphic and tabular format for simple, simultaneous cross-comparison of multiple design scenarios.

As achieving energy and IEQ objectives increase in significance, it becomes more important that the performance of fenestration systems is modeled and simulated accurately. COMFEN provides users access to a library of predefined facade components and glazing systems, including windows, shading systems, glazed wall assemblies, opaque walls, spandrels, frames, gas layers, and a comprehensive list of commercially-available glazing layers (Fig. 4.17). Each glazing data entry is based on measured data from the International Glazing Database (IGDB), a collection of data for more than 4600 glazing products, or the more recently developed Complex Glazing and Shading Database (CGDB) which adds diffusing glazings, fabrics and other shading systems. For a product to be entered to the IGDB or CGDB, manufacturers must provide measured performance data containing detailed spectral

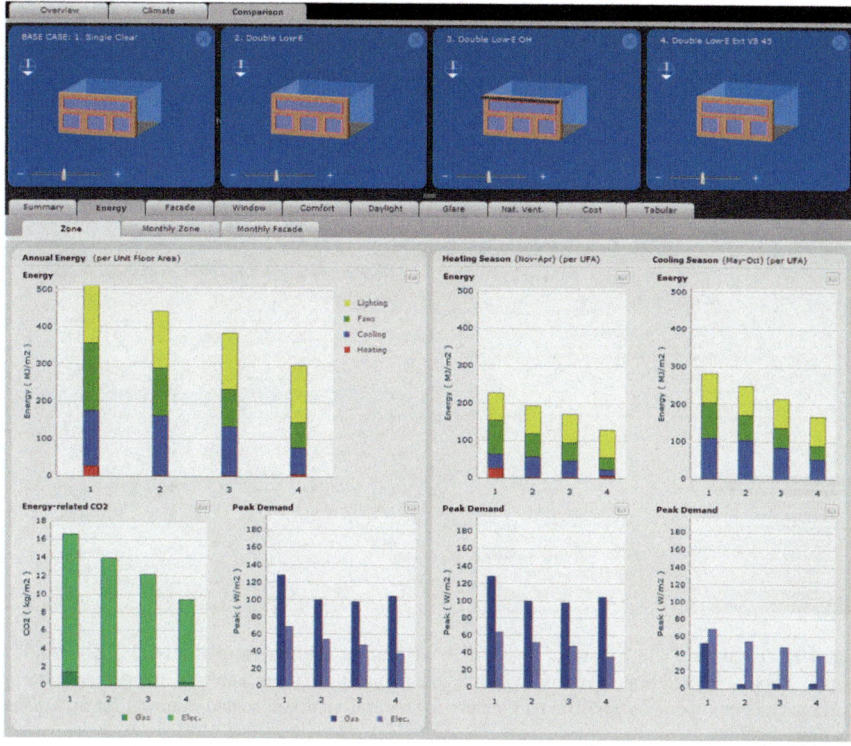

Fig. 4.16 Annual energy performance outcomes of four different fenestration scenarios for an office building located in Seattle, Washington (USA)

optical data, thermal data, structural details, and other product information. Data submitted for each product undergoes technical peer review process managed by staff at the Lawrence Berkeley National Laboratory (LBNL) to ensure that it is valid and suitable for performing accurate energy and lighting simulations.

In addition to pre-defined systems, users can build their own custom glazing systems from any combination of glazing layers and gas layers from the gas and glass library and calculate performance properties such as Tvis, SHGC, and U-factor by accessing the software tool WINDOW (LBNL 2016a, b, c) directly within COMFEN. The resulting glazing system can then be added to the glazing system library and used for current or future simulation studies.

To calculate the properties of custom systems including elements from the CGDB, such as a custom 2-layer glazing system with an internal cellular shade (Figs. 4.18 and 4.19), users must combine various glass layers, air and gas layers, and complex shading layers directly in WINDOW and save the system to their library of virtual glazing systems.

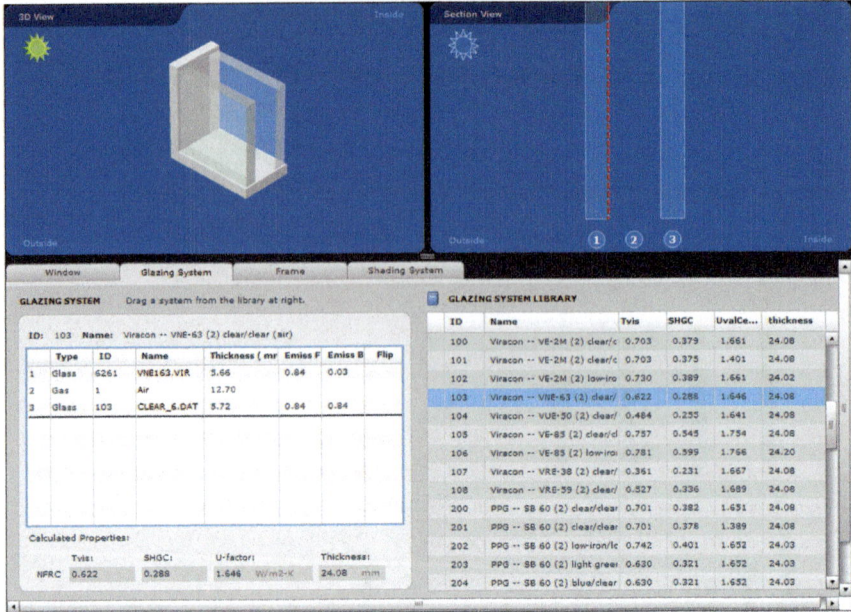

Fig. 4.17 Example "edit" view of a pre-defined glazing system from the COMFEN window system library. In addition to frame and glazing parameters, users can add interior or exterior shading systems, operable window parameters for natural ventilation, and controls for automated shading

ID	Name	ProductName	Manufacturer	Type	Material	Openness
16	WINDAT Shade with Spectral Data		Generic	Venetian (horizontal)	windat internal light	0.000
17	White Frit		Generic	Fritted glass		
19	Generic Woven Shade		Generic	Woven	Woven Shade Mate	0.050
20	Off White Blind 24 mm slat at 45 degrees		Generic	Venetian (horizontal)	White Venetian Blir	0.500
21	Dark Blue Blind 24 mm slat at 45 degrees		Generic	Venetian (horizontal)	Marine Venetian Blir	0.500
22	Clear Frit (no pigment)		Generic	Fritted glass		
23	Woven shade - 30% refl. gray, 3% openness		Generic	Woven	Woven Shade Mate	0.050
24	Woven shade - 30% refl. gray, 10% openness		Generic	Woven	Woven Shade Mate	0.050
28	Perforated screen - circular		Generic	Perforated Screen	Slat Metal A	0.500
29	1" horizontal VB (white) - 45 deg		Generic	Venetian (horizontal)	White Venetian Blir	1.000
30	3" horizontal VB (white) - 45 deg		Generic	Venetian (horizontal)	White Venetian Blir	1.000
31	3" vertical VB (white) - 0 deg (open)		Generic	Venetian (vertical)	White Venetian Blir	1.000
32	0.4" inbetween VB (white) - 45 deg		Generic	Venetian (horizontal)	White Venetian Blir	1.000
50	Cellular Shade -- Opaque -- Dark	Cellular Shade -- Opaque -- Da	Generic	BSDF		0.000
51	Cellular Shade -- Opaque -- Medium	Cellular Shade -- Opaque -- Me	Generic	BSDF		0.000
52	Cellular Shade -- Opaque -- Light	Cellular Shade -- Opaque -- Lig	Generic	BSDF		0.000
53	Cellular Shade -- Sheer -- Dark	Cellular Shade -- Sheer -- Dark	Generic	BSDF		0.000
54	Cellular Shade -- Sheer -- Medium	Cellular Shade -- Sheer -- Medi	Generic	BSDF		0.000
55	Cellular Shade -- Sheer -- Light	Cellular Shade -- Sheer -- Light	Generic	BSDF		0.000
56	Cellular Shade -- Opaque -- White Outside -- Dark Inside	Cellular Shade -- Opaque -- Wt	Generic	BSDF		0.000
57	Cellular Shade -- Opaque -- White Outside -- Medium In	Cellular Shade -- Opaque -- Wt	Generic	BSDF		0.000
58	Cell in Cell Cellular Shade	Cellular Shade -- Opaque -- Wt	Manufacturer	BSDF		0.000

Fig. 4.18 Browsing for cellular shade in the Complex Glazing and Shading Database [CGDB (http://windowoptics.lbl.gov/data/cgdb)]

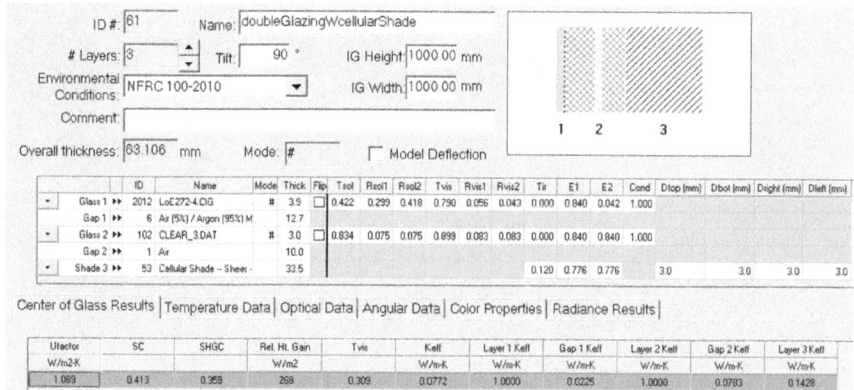

Fig. 4.19 Use of the software tool WINDOW to generate a custom complex fenestration system, calculate material properties, and save to the glazing system library for import to COMFEN for use in performance analysis. The above system consists of 5 layers (order is from exterior to interior): *1* glass with a low-e coating, *2* an air/argon gas layer, *3* a clear glass lite, *4* an air gap, *5* a cellular shade

COMFEN allows extensive and rapid exploration of energy properties at the annual, monthly and/or hourly level to gain further insights into performance levels. While annual EUIs are critical high level performance metrics it is often important to understand the energy use and comfort cycles during certain months or hourly values across a day. As an early design tool, COMFEN emphasizes comparative performance across alternatives and the user interface is set up to facilitate visual comparisons of those options.

The extensive analysis options via "virtualization" in tools like COMFEN, with built in libraries of key fenestrations components, paired with easy-to-use interfaces linked to powerful, validated simulation engines, enables physically accurate, climate-based testing and evaluation in a simulation environment, which is much faster and cheaper than testing in a physical testbed.

4.3.5 Humans

Even in the most sophisticated simulation tools and workflows, the presence and environmental preferences of occupant are still often represented by simplistic, static and universally applied assumptions. In practice, simplistic application of human factors data limits the energy and carbon reduction potential of energy efficiency measures, and can lead to operational challenges (e.g. Fig. 4.20) and discrepancies between anticipated and measured energy consumption or IEQ performance objectives. Although it is unrealistic to assume that the preferences and behaviors of a specific population of building occupants can be routinely predicted

Fig. 4.20 Interior roller shades deployed in perimeter zone of daylit office building, limiting daylight transmission and views, and leading to the use of electrical ambient lighting adjacent to windows during daytime

with a high degree of accuracy, (particularly prior to construction of the project), it is important for designers to be aware of the large array of human-factors assumptions embedded in software-based design tools and understand the impacts these assumptions may have on anticipated performance outcomes.

These assumptions fall into two general categories: occupant behaviors and occupant preferences. In the behaviors category, assumptions include schedules of occupancy, manual control of electrical ambient and task lighting, and manual control of shading devices. Occupant preferences (discussed in Chap. 2) are represented in criteria for horizontal daylight illuminance, and the solar and lighting conditions associated with glare and triggers for the deployment of shading devices. Not surprisingly these assumptions are interrelated and can be assumed to differ substantially across a population of occupants based on many factors. Where knowledge exists for predicting occupant outcomes, the designer should try to verify that the underlying assumptions are relevant to that project. For example, existing assumptions for the conditions that drive occupant use of interior shading devices are largely derived from cellular, single-occupancy offices, where occupants can more reliably be expected to operate blinds and lights at least to reduce glare and discomfort, although not necessarily to optimize energy use. This knowledge may be problematic when applied directly to predict behaviors in much larger open-plan offices, where occupants share control over shading devices. Even in the application of automated systems, the design team should have a basic evidence-based knowledge of appropriate human-factors performance boundaries (as discussed in the following section).

As designers increasingly rely on software-based simulation workflows to gain confidence in the energy and IEQ performance of a project prior to construction, awareness of the limits of existing approaches to modeling human factors takes on increasing significance. The emergence of occupant-aware, "human-in-the-loop" control schemes (discussed in Chap. 3), and scalable Post Occupancy Evaluation (POE) methodologies (discussed in Chap. 6) present promising mechanisms for improving the representation of occupants in energy and lighting simulation workflows by increasing and systematizing the collection of occupant data from projects in use. However, occupants remain a key source of variability and thus uncertainty in the outcomes from simulation-based daylighting design.

4.3.5.1 Modeling Occupant Behavior

Occupant control of interior shading devices has a significant impact on daylight transmission and view, and adds additional complexity and uncertainty to the simulation process. In many cases, the modeling assumptions for occupant behavior (such as the illuminance thresholds conventionally used for triggering opening and closing of shades) can have a greater impact on simulation outcomes than building geometry, material attributes, or climate. As one example, in a simulation-based annual study of three lighting and shade control patterns (automated, active user, static user), the reported energy savings ranged from 0 to 60% (Reinhart 2004).

A recent review of existing research on occupant control of shading devices in office buildings reveals a general consensus for the hypothesis that shading devices are deployed by occupants to control glare, direct sun penetration and overheating (Van Den Wymelenberg 2012). Regarding the frequency of shade operation, there are two general hypotheses. The first assumes that occupants deploy shading devices in response to the magnitude of solar radiation incident on the workspace and retract shading devices on a daily basis (either the following day, or when the stimulus no longer exceeds the threshold for deployment). This "active operator" hypothesis is often adopted in computer simulations of daylight availability particularly when the goal is to estimate performance potentials (Lee and Selkowitz 1995; Reinhart 2004; Heschong et al. 2010). The second, "worst case scenario," hypothesis emerges from observational studies of buildings in use (Rubin et al. 1978; Rea 1984; Foster and Oreszczyn 2011; Inkarojrit 2005) and is based on the conclusion that occupants appear to position shading devices according to the "worst case" solar control condition based on perceptions formed over weeks or months, and rarely adjust them. Whether occupants behave as "active operators" or position shades for "worst case" solar control conditions has a significant effect on daylight availability, visual connection to the outdoors, and the potential for electrical lighting energy reduction. Any given population of occupants is likely to have a mix of these behavior types. In addition evolving shading hardware and software solutions, and changes in interior office designs might enhance or detract from the viability of existing design solutions.

Due largely to the lack of consensus for appropriate models of occupant behavior, shade operation is poorly defined in existing approaches to performance assessment. For example, the modeling requirements for compliance with ASHRAE 189.1 and California Title-24 Section 6 do not address the issue of shade operation whatsoever in the prediction of electrical lighting reduction from photocontrols. The Illuminating Engineering Society (IES) *Approved Method: IES Spatial Daylight Autonomy (sDA) and Annual Sunlight Exposure (ASE)*, commonly referred to as LM-83 and incorporated into the LEEDv4 Daylight credit compliance procedure, is the first standard to include a protocol for modeling occupant shade control. It assumes an active user, such that shades are fully lowered when over 2% of the floor area receives direct sun (as indicated by interior horizontal workplane illuminances grater than 1000 lx), and fully retracted when this criteria is not met. While this is a plausible basis for design it is not yet supported by extensive field data.

Since occupants are fallible, automated shading systems can be applied to projects with the goal of increasing the reliability by removing occupants from the management of daylight transmission. This approach requires a design solution that delivers conditions that are acceptable when shades are retracted, and that adequately manage solar and glare when shades are deployed. Thus, both the control parameters (e.g. indicators and thresholds for deployment) and fenestration components (e.g. glazing visual light transmittance, roller shade fabric openness factor, sensors) must be thoughtfully considered and based on knowledge of end-user preferences or acceptance levels. Even small variations within these parameter assumptions can have significant impacts on annual daylight availability and, if poorly addressed, trigger permanent retrofits to ameliorate conditions for occupants (Konis 2012).

The following figures present an example applying Climate Based Daylight Modeling (CBDM) to a hypothetical 10 m-wide section of a side-lit building floor plate with a south-east facing solar orientation located in San Francisco, California (Fig. 4.21). The figures sequentially illustrate the assumed and observed impact of occupants on design intent and daylighting performance. An example space is modeled as a 10 m-wide by 12 m-deep zone, half the depth of the floor plate (24 m) and 3 m in interior height, with typical interior surface reflectances (floor = 0.3, walls = 0.5, ceiling = 0.5). No interior furnishings are included in the model. Local meteorological data for downtown San Francisco are used to simulate the annual Daylight Autonomy (DA_{300}) results across a hypothetical workplane.

The first simulation outcome (Fig. 4.22) shows the daylighting potential of the perimeter zone prior to the addition of facade glazing, and indicates that over 90% of the zone requires no supplemental electrical lighting in an ideal case. The second simulation includes the addition of a fully-glazed facade, using high-performance glazing with typical high visible light transmittance of (Tvis = 0.60) for a glazing

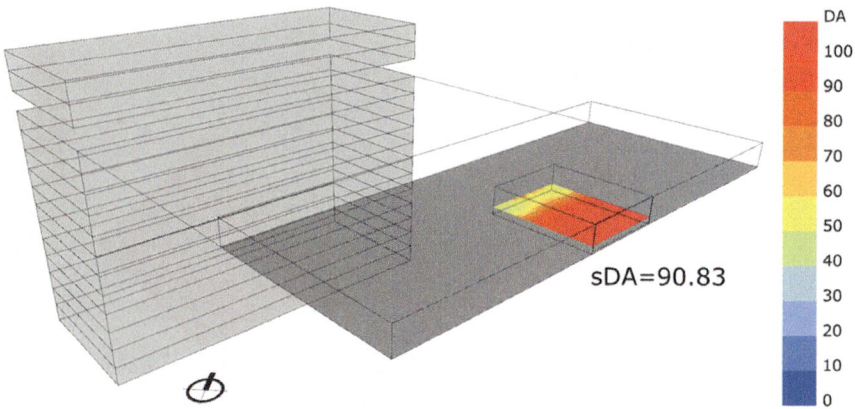

Fig. 4.21 Example SE-facing sidelit zone occupying half the depth of a 24 m wide floor plate of a commercial office building located in San Francisco, California. Simulation results show the daylighting potential of the zone prior to the addition of facade glazing

Fig. 4.22 Simulation 1: model with floor-to-ceiling facade glazing … annual daylighting potential (sDA) of the zone prior to the addition of facade glazing

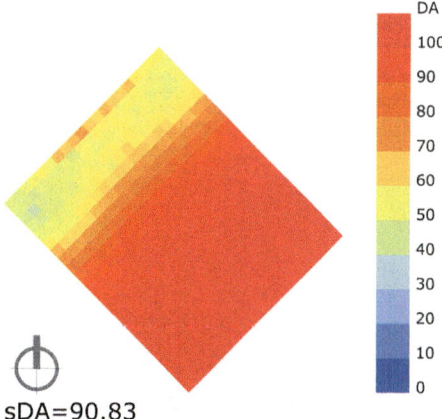

that has excellent solar control (e.g. SHGC = 0.30) (Fig. 4.23). The third simulation includes the addition of an exterior perforated metal solar control screen (50% openness at normal incidence). This third simulation outcome (Fig. 4.24) reflects the need (even in moderate climates) for projects targeting low-energy goals to supplement large areas of facade glazing with exterior attachments to reduce peak solar heat gains. However, it is important to note that exterior solar control screens are often also considered sufficient for providing glare and solar control to occupants working in perimeter zones.

The fourth simulation includes the addition of manually operated interior roller shades (3% openness factor) controlled using the theoretically derived "active operator" shade control model. Similar to previous examples, the environmental analysis plugin Honeybee (Roudsari and Pak 2013) is used to build a custom

Fig. 4.23 Simulation 2:
model with floor-to-ceiling
facade glazing (Tvis = 0.6)

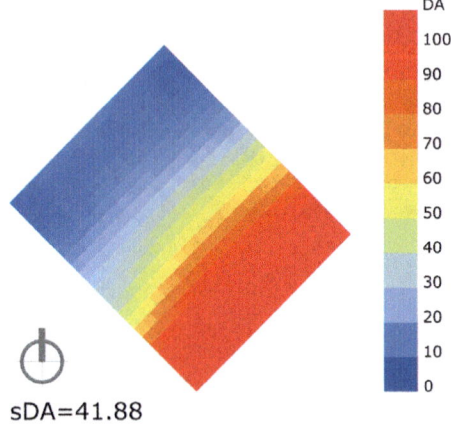

sDA=62.50

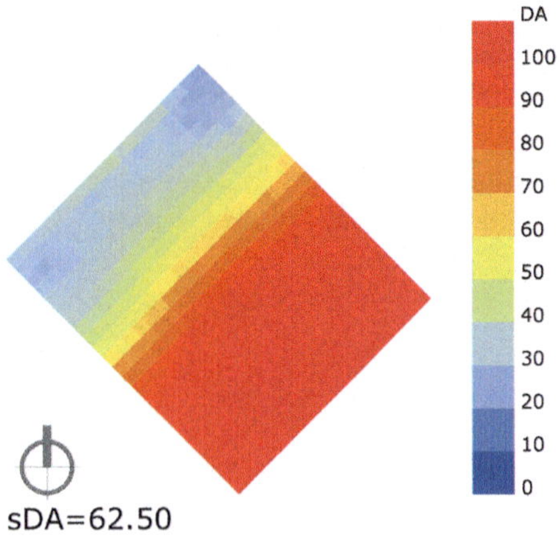

Fig. 4.24 Simulation 3:
original design intent. Model
with floor-to-ceiling facade
glazing (Tvis = 0.6) and
exterior metal scrim
(0.5-openness at normal
incidence)

sDA=41.88

Fig. 4.25 Simulation 4:
model with floor-to-ceiling
facade glazing (Tvis = 0.6),
exterior metal scrim
(0.5-openness at normal
incidence) and interior roller
shade retrofit (0.03 openness
factor) operated with a
2000 lx deployment threshold

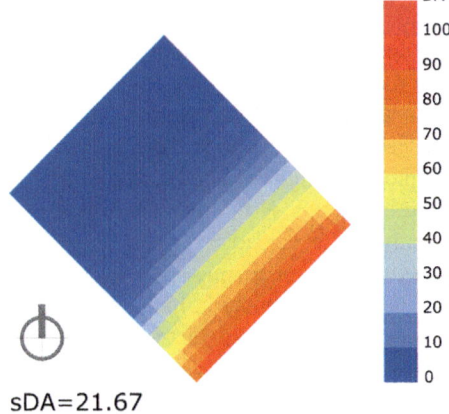

sDA=21.67

workflow to access the Radiance-based DAYSIM daylight analysis software (Reinhart 2016). DAYSIM allows designers to model spaces with dynamic shading systems, for example, venetian blinds, roller shades, or electrochromic glass. As shown in Fig. 4.27, DAYSIM was used to generate annual illuminance profiles resulting from the application of a simple deterministic model (Reinhart 2004) of occupant shade control behavior. In this case, model parameters are based on the assumption that shades will be lowered when interior horizontal illuminances

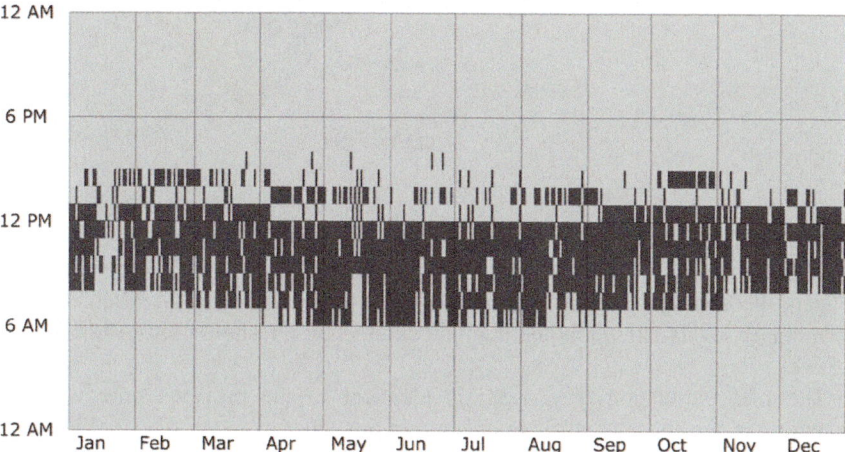

Fig. 4.26 Shades use profile for occupant shade control model implemented in Fig. 4.25. *Dark grey* indicates deployment of shades

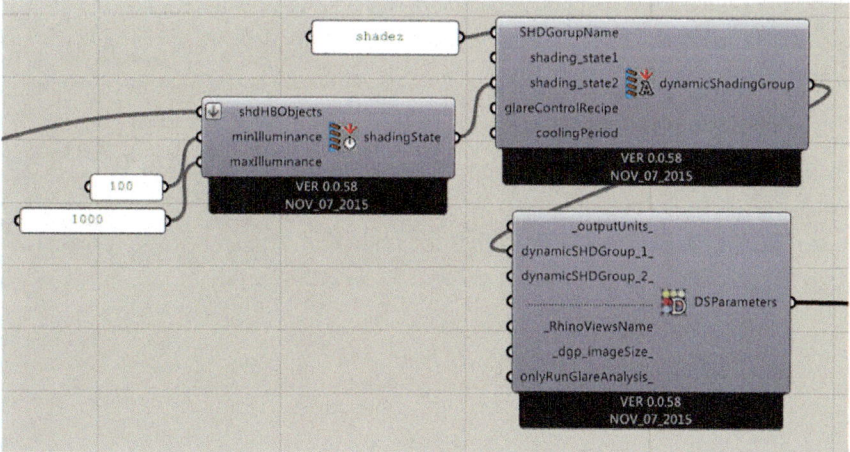

Fig. 4.27 Screen capture from visual scripting workflow showing view of Honeybee components which prepare shade control parameters for DAYSIM

Fig. 4.28 Simulation 5:
Model with floor-to-ceiling
facade glazing (Tvis = 0.60),
exterior metal scrim
(0.5-openness at normal
incidence), and interior roller
shade retrofit (0.03 openness
factor) operated with a
1000 lx deployment threshold

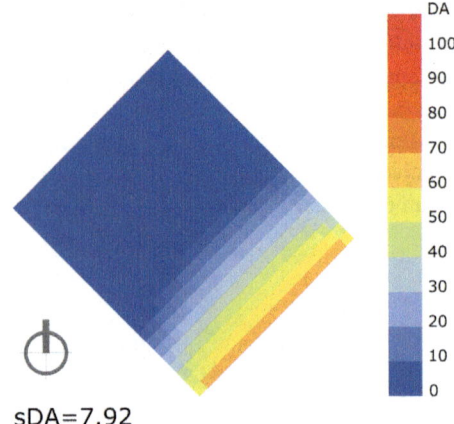

sDA=7.92

exceed 2000 lx (the upper limit of the Useful Daylight Illuminance (UDI) metric), and will be raised again when interior illuminances fall below 300 lx (a setting that reflects the assumption that occupants will raise the shade again when daylight illuminance levels fall below the daylight sufficiency threshold used in LM-83 and LEED.

The fifth simulation (Fig. 4.28) is identical to the previous one, with the exception of the shade control parameters. In this latter case, the shades (Fig. 4.29) are deployed when interior daylight illuminances on the workplane exceed 1000 lx, and retracted when illuminances fall below 100 lx, this second parameter value

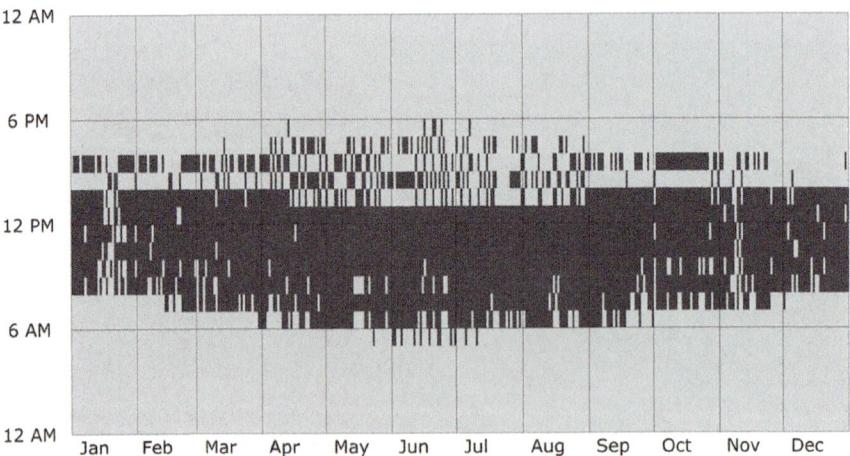

Fig. 4.29 Shades use profile for more restrictive occupant shade control model implemented in Fig. 4.28. *Dark grey* indicates deployment of shades

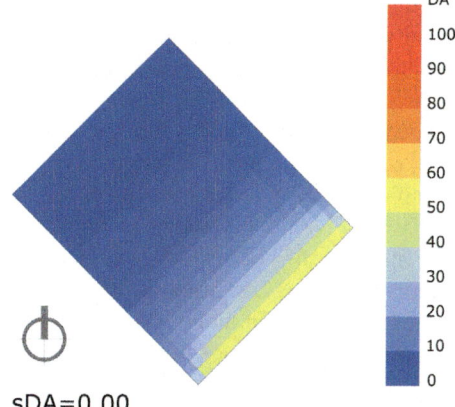

Fig. 4.30 Simulation 6: model with floor-to-ceiling facade glazing (Tvis = 0.60), exterior metal scrim (0.5-openness at normal incidence), interior roller shade retrofit (0.03 openness factor) operated with a 1000 lx deployment threshold, and an interior solar control film (Tvis = 0.24) added to address issues of glare and solar overheating in the perimeter zone

sDA=0.00

reflects the tendency of occupants to retract shades at a time when the source of discomfort is no longer present. For the final simulation (Fig. 4.30), a solar control film (Tvis = 0.24) is added to the glazing to reflect the observed outcome where a permanent facade retrofit was made to address issues of glare and solar overheating along the perimeter. The contrast in spatial Daylight Autonomy (sDA) outcomes between design intent and performance in use illustrates the significance of occupant intervention in daylighting effectiveness both in terms of dynamic and permanent changes to facade light transmission.

4.3.5.2 Discussion

Not surprisingly, the simulation case study presented in Figs. 4.21, 4.22, 4.23, 4.24, 4.25, 4.26, 4.27, 4.28, 4.29 and 4.30 demonstrates via modeling that various physical design options with glazing and shading as well as assumptions for occupant visual comfort requirements and shade control behavior can have significant impacts on simulation outcomes for daylight availability. For example, the original design assumed that an exterior solar control screen would be sufficient for glare control and that no interior shading devices needed to be installed. It should be noted that the current compliance procedures for the LEED Daylighting EQ credit allow designers to exclude interior shading from simulation-based compliance models if the devices are not planned for installation in the project. In this case study, the retrofit addition of interior shading devices to improve control over glare, when modeled with a relatively high shade deployment threshold (2000 lx), was found to reduce sDA from 42 to 22%, resulting in a 48% reduction in daylight availability from the original design intent. When the even more conservative shade deployment threshold (1000 lx) specified in LM-83 and LEED was applied, daylight availability was further reduced by 81%, from an sDA of 42 to 8%. Finally, the retrofit addition of a solar control film, combined with shades resulted in no regions of the space achieving the Daylight Autonomy criteria (300 lx daylight illuminance over 50% of occupied hours annually), or an sDA of zero (0). This outcome matches closely with the observed outcome (see Fig. 4.20), and is particularly notable given the potential of the zone, which can be considered almost completely daylight autonomous in the ideal case.

To improve fidelity between simulation models and projects in use, it is critical for models of occupant behavior to become an integral component of simulation-based design workflows. It is equally critical that models be informed through observation and evaluation of human behavior in buildings, and that this information is fed back into the design process, not simply to improve model fidelity, but to drive innovation in the design and operation of the daylighting strategy. In addition to the significant variations in people that can be expected, one can also expect significant variations in shade control behaviors with facades that are designed to offer more complex shading configuration possibilities to occupants. It is important to recall that existing behavioral models were primarily developed from studies conducted in single occupancy offices with venetian blinds where a single occupant is assumed to operate a single blind (or, where multiple windows exist, occlude all windows to the same level). In application to an open plan office where the facade is subdivided vertically into a row of vision windows and multiple (e.g. 2) rows of upper clerestory windows, (such as the facade shown in Fig. 3.50 of Chap. 3), no guidance exists for how predicted changes in overall occlusion level are represented among the multiple shading devices available for a given occupant.

The topic of field evaluation and feedback is discussed in detail along with emerging monitoring approaches in the following chapter (Chap. 6). As discussed in Chap. 3, the growing interest in IoT-sensing for human-in-the-loop control and post-occupancy evaluation may help enable physical data to be collected continuously in real time data on virtually all conditions in the building, at a granular level related to workspaces. However, the task remains to place these data in context with end-user requirements and preferences, some of which are highly subjective and may be collected much less frequently through various POE mechanisms, user-overrides to automated controls, or observed through intelligent (i.e. IoT-enabled) manually controlled systems. Aggregating across multiple occupants and multiple projects, these data can be leveraged to validate or adapt existing human-factors models and then be applied to develop improved models that may provide the performance criteria needed to routinely design comfortable daylit buildings (architectural forms and materials) and to define optimal control strategies for automated systems. The future of smart automated systems likely rests on three premises: (1) that sensor and control hardware and software become better and cheaper; (2) that systems based on that hardware/software can be installed, commissioned and operated reliably over time; and (3) that we understand human factors comfort and performance needs and behavioral responses sufficiently to take advantage of (1) and (2). The inexorable progress with technology and software provides confidence in the first two but a concerted effort needs to be made to address the human-in-the-loop element to drive overall success in buildings.

4.4 Lessons and Feedback from the Built Environment

While annual, climate based thermal and lighting simulation workflows significantly improve designer's ability to assess and refine building performance during design, there are many real-world considerations that are poorly accounted for in software-based building design, and can be more effectively addressed by experimentation in the physical world. These considerations generally fall into three general overlapping categories: (1) systems integration, (2) human factors, and (3) operational reliability.

While the application of new building technologies often focuses on the performance of a single piece of technology, it is the overall performance of multiple technologies working as an integrated system that will better address building needs and drive innovation in the future. This is particularly important to consider for low-energy perimeter zone systems, where active use of the building envelope (e.g. solar control daylighting, natural ventilation, and charging/discharging thermal mass) paired with controllable lighting and HVAC systems has great potential for achieving low and zero net energy performance objectives. But while these integrated solutions may be more technically challenging to optimize and implement, these solutions can also improve project economics as savings from some systems can be used to offset increased costs in other areas. For example, improved external solar shading, with a natural ventilation option and some night time thermal storage might allow a mechanical cooling system to be eliminated from the design or greatly downsized, with the associated cost savings used to offset the costs of the enhanced facade system.

The behavior of occupants is invariably complicated, and more so when they move to spaces with new design features and controls. Recognizing that manual controls are not consistently used by most occupants, architects and engineers may elect to design occupants out-of-the-loop and rely on sensors and algorithms to manage all operable systems. But these systems rarely work equally well for all occupants and if proper overrides are not accessible to users, they may take actions that further reduce system performance in ways unanticipated by designers and their software-based controls.

Finally the reliability and consistency of building operations in the real world is hard to predict in software, since it involves capturing construction modifications to design, other new as-built details, the imperfections of commissioning of complex controls and finally the actions of occupants. This is why measured building performance with respect to energy and comfort so often deviates from design expectations based on even the most extensive simulation studies.

In the long term, simulation tools and new design processes will address more of these issues. Post occupancy studies in completed buildings will also fill in some of the "reality" knowledge gaps but this data collection and evaluation can be slow and costly using conventional approaches. In the short term examining the performance of a collection of promising energy-efficient technologies under real operating conditions and in real time using a full-scale test facility has the potential to reduce the gap between design expectations and performance in use. A full scale test bed, properly configured and instrumented, can fill gaps that simulation tools will miss, at much lower cost and in much less time than instrumented studies of completed buildings. In regard to human-factors, one significant benefit of a full-scale physical space is the potential for direct human evaluation of the Indoor Environmental Quality (IEQ) factors resulting from the integrated performance of various components in use under real sun and sky conditions. This is particularly important for factors such as view and glare, which are difficult to assess in a simulation environment. Finally, in regard to operational reliability, simulation-based tools normally assume idealized operation of technology since actual operation is unknown, and available virtual models may not capture the details of specific commercially available product. Physical testbeds offer the ability to evaluate the performance of promising new technologies over time under dynamic operating conditions, where valuable insights into the state of the technology can be gained, for example, the switching speed of electrochromic glazing in cold weather, or the positional precision of an automated light redirecting venetian blind after 6-months of dynamic adjustments in slat position.

4.4.1 Mock-Ups

The concept of exploring a future building space in a virtual reality software-based "mockup" is gaining traction and interest. Physical mockups still possess tremendous predictive power and usefulness as well if implemented strategically. The building industry often builds "look and feel" mockups for visual inspection and other physical mockups for certification testing, e.g. facade mockups for air, water and structural testing. But it rarely builds "performance mockups" that would allow design and engineering details to be fully worked out before specifications are completed and construction begins. Working closely with a motivated building owner, the LBNL facade team had the opportunity to utilize a large full size mockup (~ 500 m^2) of the New York Times Headquarters building in New York City to define, test and optimize automated shading and dimmable lighting systems with furniture, fixed shading, carpeting, ceilings, etc. (Lee et al. 2005). The tests were conducted over an 18-month period with multiple equipment suppliers and extensive instrumentation. Although the mockup cost over US$1M, the owners documented construction cost savings alone (e.g. fewer change orders) exceeded the mockup construction cost due to savings in the details of the steelwork design that were resolved before the 52 story building was under construction. This extensive mockup testing of the New York Times headquarters design was carefully documented and followed by a heavily instrumented post occupancy evaluation of the building several years after occupancy that showed large energy savings and high levels of occupant satisfaction (Lee et al. 2013).

The Times mockup was custom designed for that single building and could not be reused. LBNL has built and operates a series of reconfigurable testbeds to carry out different facade mockup studies at several scales. Three side-by-side office sized test rooms, south facing are shown in Fig. 4.31. The facility has been used for engineering studies and for human factors studies as well in an occupied mode. The Mobile Thermal Test Facility (MoWiTT) is a dual chamber, high accuracy facility used for facade component testing in different climates and orientations. (Klems 1988). By its design it tests smaller facades and its extensive interior instrumentation makes it impossible to have occupants work on the indoor space.

LBNL's latest testbed, the Facility for Low Energy Experiments in Buildings (FLEXLAB), completed in 2013 has three south-facing modules about 12.2 m (40′) wide and 9.1 m (30′) deep, each with a side-by-side sets of rooms, both one story and two, and one set of side-by-side rooms (Fig. 4.32) that can rotate to any orientation (Fig. 4.33) (LBNL 2016a, b, c).

Fig. 4.31 LBNL windows testbed facility (south-facing facade) showing comparison of two exterior shading systems to an unshaded window reference case

Fig. 4.32 Exterior view of the U.S. Department of Energy's Facility for Low Energy Experiments in Buildings (FLEXLAB) facility at LBNL showing side-by-side rotational test building. *Image credit* LBNL

These facilities can be used for engineering tests on virtually all types of envelope/shading/daylighting, any lighting system, and all types of air or water based HVAC, ceiling or floor located. The rooms can also be occupied for human factors studies and include a number of sensors configured for comfort measurements (Figs. 4.34 and 4.35). These include horizontal illuminance, air temperature and relative humidity, mean radiant temperature, supply air temperature and flow, infrared measurement of surface temperature, air velocity, discomfort glare (e.g. luminance maps and DGP calculations), and blind position. Luminance maps can also be examined to assess factors related to view. A more detailed description of glare sensing is provided in Chap. 2. Side-by-side testing is particularly important for rigorous comparative testing of design or operational alternatives since weather conditions are always changing and test data collected sequentially over time is often hard to normalize and compare.

    LBNL has recently carried out testbed studies in a mockup of a new building design with a large glazed facade but with carefully selected high performance glass, external fixed shading and automated roller shades for glare and thermal

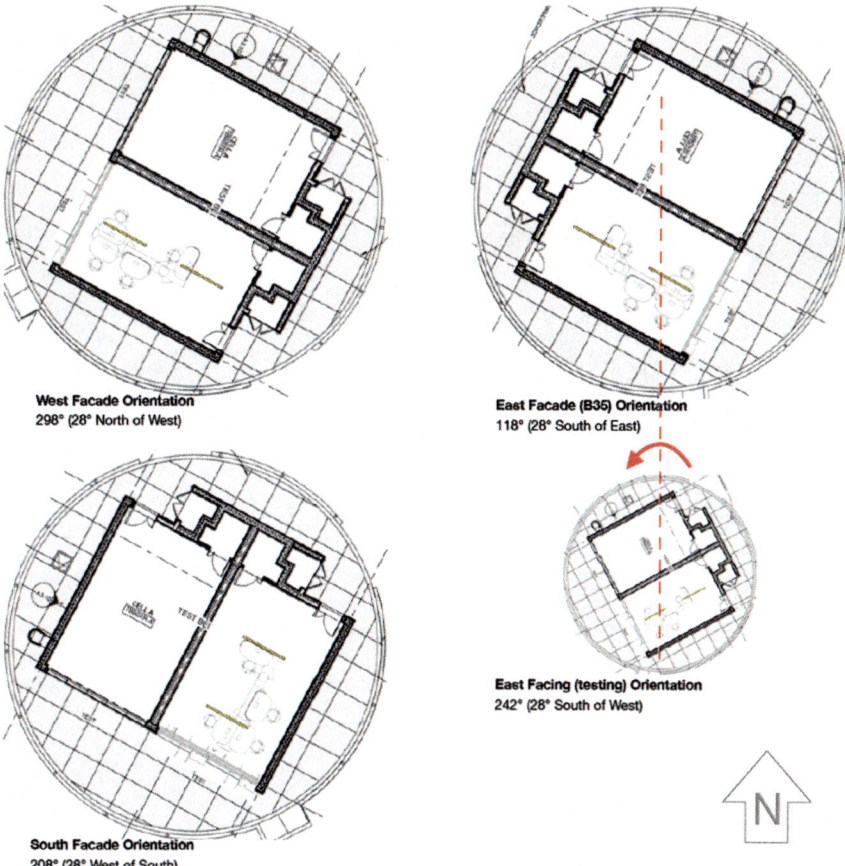

West Facade Orientation
298° (28° North of West)

East Facade (B35) Orientation
118° (28° South of East)

East Facing (testing) Orientation
242° (28° South of West)

South Facade Orientation
208° (28° West of South)

N

Fig. 4.33 Example of test cell orientations used to evaluate building facade and automated lighting and shade control systems. *Image credit* LBNL

comfort management (McNeil et al. 2014). This study focused on automated control that delivered comfort and occupant response as much as energy optimization. Tests in the rotating facility were used to optimize solutions for both south and west orientations, each with different fixed external shading conditions. Results from FLEXLAB demonstrated that additional high quality floor space could be "recaptured" if glazing with fixed and active shading are thoughtfully designed and operated. The study was launched late in the design process when structural steel was already being erected but the owner and contractor were able to use the results for final selection of lighting fixtures and controls, shading fabric and operating controls, and furniture location.

Fig. 4.34 Interior view of the U.S. Department of Energy's Facility for Low Energy Experiments in Buildings (FLEXLAB) showing Indoor Environmental Quality (IEQ) monitoring equipment including installation of shielded thermocouples at regular heights to assess thermal stratification. *Image credit* LBNL

Fig. 4.35 Comfort sensing equipment installed in FLEXLAB: *1* Licor illuminance sensor for calibrating HDR images taken by *2* DSLR camera with 180° lens, *3* air velocity sensor, *4* mean radiant globe temperature sensor, and *5* shielded temperature sensor. *Image credit* LBNL

Acknowledgements The authors would like to acknowledge the contribution of Alejandro Gamas who developed the grasshopper workflow documented in Sect. 4.3.3.

References

American Institute of Architects (AIA) (2014) AIA 2030 commitment 2014 progress report. http://www.aia.org/aiaucmp/groups/aia/documents/pdf/aiab107447.pdf. Accessed 06 Sept 2016

American Society of Heating and Refrigeration Engineers (ASHRAE) (2013) ANSI/ASHRAE/US GBC/IES Standard 90.1-2013. American Society of Heating and Refrigeration Engineers, Atlanta, GA

Andersen M, et al (2008) An intuitive daylighting performance analysis and optimization approach. Build Res Inf 36.6: 593–607. ©2008 Taylor & Francis

Ashdown I, Ward G (2013) http://www.radiance-online.org/learning/documentation/manual-pages/pdfs/gendaymtx.pdf

AMC Bridge (2016) CAD to Earth. http://amcbridge.com/labs/cloud-apps/cadtoearth. Last accessed 2 Sept 2016

Caldas L (2008) Generation of energy-efficient architecture solutions applying GENE_ARCH: an evolution-based generative design system. Adv Eng Inf 22(1):59–70. ISSN 1474-0346. doi:10.1016/j.aei.2007.08.012

Compagnon R, Raydan D (2000) Irradiance and illuminance distributions in urban areas. In: Proceedings PLEA 2000, Cambridge UK, July 2000, pp 436–441

Deru et al (2011) U.S. Department of Energy Commercial Reference Building Models of the National Building Stock. Technical Report. NREL/TP -5500-46861

Foster M, Oreszczyn T (2011) Occupant control of passive systems: the use of venetian blinds. Build Environ 36(2001):149–155

Gagne J, Andersen M (2012) A generative facade design method based on daylighting performance goals. J Build Perform Simul 5.3:141–154

Heschong L., Saxena M, Higa R (2010) Improving prediction of daylighting performance. In: Proceedings of the ACEEE 2010 summer study on energy efficiency in buildings. American Council for an Energy-Efficient Economy, Washington, D.C.

Illuminating Engineering Society (IES) (2012) LM-83. Approved method: IES spatial daylight autonomy (sDA) and annual sunlight exposure (ASE). ISBN 978-0-87995-272-3

Inkarojrit V (2005) Balancing comfort: occupants' control of window blinds in private offices. Doctoral Dissertation. University of California, Berkeley

Klems JH (1988) Measurement of fenestration net energy performance: considerations leading to development of the mobile window thermal test (MoWiTT) facility. J Sol Energy Eng 110 (1988):208–216

Konis KS (2012) Effective daylighting: evaluating daylighting performance in the san francisco federal building from the perspective of building occupants. Center for the Built Environment, UC Berkeley. Retrieved from: http://escholarship.org/uc/item/7qg1945w

Konis K, Gamas A, Kensek K (2016) Passive performance and building form: an optimization framework for early-stage design support. Sol Energy 125:161–179

LBNL (2016a) COMFEN. https://windows.lbl.gov/software/comfen/comfen.html. Last accessed 2 Sept 2016

LBNL (2016b) WINDOW. https://windows.lbl.gov/software/window/window.html. Last accessed 2 Sept 2016

LBNL (2016c) FLEXLAB. https://flexlab.lbl.gov/. Accessed 6 Sept 2016

Lee ES, Selkowitz SE (1995) The design and evaluation of integrated envelope and lighting control strategies for commercial buildings. ASHRAE Trans Chicago, Il 101(1):326–342 (28 Jan–1 Feb)

Lee ES, Selkowitz S, Hughes G, Clear R, Ward G, Mardaljevic J, Lai J, Inanici M, Inkarojrit V (2005) Daylighting the New York Times Headquarters Building: Final Report. 2005. Lawrence Berkeley National Laboratory, Berkeley, CA. LBNL-57602

Lee ES, Fernandes LL, Coffey B, McNeil A, Clear R, Webster T, Bauman F, Dickerhoff D, Heinzerling D, Hoyt T (2013) A post-occupancy monitored evaluation of the dimmable lighting, automated shading, and underfloor air distribution system in The New York Times Building. LBNL Technical report, January 2013

Mardaljevic J, Rylatt M (2000) An image based analysis of solar radiation for urban settings. In: Proceedings PLEA 2000, Cambridge UK, July 2000, pp 442–447

McNeil A, Kohler C, Lee ES, Selkowitz S (2014) High performance building mockup in FLEXLAB. LBNL report number 1005151. http://eetd.lbl.gov/publications/high-performance-building-mockup-in-f

Perez R, Seals R, Michalsky J (1993) All-weather model for sky luminance distribution—preliminary configuration and validation. Sol Energy 50(3):235–243

Rea M (1984) Window blind occlusion: a pilot study. Build Environ 19(2):133–137

Reinhart CF (2004) Lightswitch-2002: a model for manual and automated control of electric lighting and blinds. Sol Energy 77(1):15–28

Reinhart C (2016) Daysim advanced daylight simulation software. http://daysim.ning.com/. Last accessed 2 Sept 2016

Robinson D, Stone A (2004) Irradiation modelling made simple: the cumulative sky approach and its applications. In: The 21st conference on passive and low energy architecture. Eindhoven, The Netherlands, 19–22 September 2004

Roudsari MS, Pak M (2013) Ladybug: a parametric environmental plugin for grasshopper to help designers create an environmentally-conscious design. In: Proceedings of the 13th international IBPSA conference held in Lyon, France Aug 25–30th

Rubin AI, Collins BL, Tibbott RL (1978) Window blinds as a potential energy saver—a case study (NBS Building Science Series 112). U.S. Department of Commerce, National Bureau of Standards, Washington, DC

Tregenza P, Sharples S (1993) Daylighting algorithms. ETSU S 1350-1993, UK

Van Den Wymelenberg K (2012) Patterns of occupant interaction with window blinds: a literature review. Energy Build 51:165–176

Vier R (2016) Octopus plug-in for grasshopper. http://www.grasshopper3d.com/group/octopus. Last accessed 2 Sept 2016

Ward G, Shakespeare RA (1998) Rendering with radiance. Morgan Kaufmann Publishers, 664 pp

Chapter 5
Case Studies

5.1 Introduction

The case studies in this chapter present large-scale, commercially-driven projects targeting low and Zero Net Energy (ZNE) located across a range of climates and contextual conditions. Although they constitute an extremely small sub-set of daylit buildings, they are representative of the extent to which practice has begun to address the primary themes outlined in this book. All six examples involve the application of daylighting within a low-energy building concept driven by specific energy and Indoor Environmental Quality (IEQ) performance targets established early in design. Notably, these projects represent a departure from standard practices of deep floor-plate, sealed envelopes, unshaded fenestration, and cellular interiors to more environmentally responsive building forms and envelope systems paired with open-plan office environments that permit greater access to daylight, views and natural ventilation for occupants. Through effective solar control and high-efficiency building envelopes, all six projects are capable of meeting heating and cooling needs without the use of a forced-air heating and air conditioning system. Underlying the design strategies and technologies applied in these projects, (such as photocontrolled individually addressable electrical lighting systems, environmentally-aware automated facade systems, and complex fenestration), are novel collaborative processes and design workflows that integrate performance-based feedback to inform design decisions. Two case studies (the RSF and the NYT) present promising examples of the benefits of performance-based project delivery, detailed commissioning practices, and Post Occupancy Evaluation (POE) to evaluate the fidelity of daylight design intent and model predictions with measured energy data and occupant feedback.

© Springer International Publishing Switzerland 2017
K. Konis and S. Selkowitz, *Effective Daylighting with High-Performance Facades*, Green Energy and Technology, DOI 10.1007/978-3-319-39463-3_5

The following sections present case studies of the following projects: (5.2) the John and Frances Angelos Law Center, designed by Behnisch Architekten, (5.3) NewActon Nishi, designed by Fender Katsalidis, (5.4) the Research Support Facility (RSF), designed by RNL, (5.5) the Bullitt Center, designed by the MillerHull Partnership, (5.6) the New York Times Headquarters (NYT), designed by Renzo Piano, and (5.7) the Nordea Bank Headquarters, designed by Henning Larsen Architects.

5.2   John and Frances Angelos Law Center

The John and Frances Angelos Law Center (Fig. 5.1) demonstrates the integration of building form, varying program elements, and facade systems to minimize demand for mechanical space conditioning and electrical lighting energy in a large 17,837 m^2 (192,000 ft^2) academic building. Located in a cooling-dominated climate, where sealed facades and air-conditioning are standard practice, the project presents an environmentally-responsive alternative model with potential co-benefits to building occupants through the provision of greater access to daylight, visual connection to the exterior, and control over indoor environmental conditions. Designed with the objective of achieving a LEED Platinum rating, the project is predicted to achieve a site EUI of 125 kWh/m^2-year (40 kBtu/ft^2-year) annually. If this performance outcome where to be achieved, the project would meet the energy target of the AIA's 2030 Commitment with a 62.2% carbon emission reduction compared to the Energy Star 50th percentile building (Table 5.1).

5.2.1   Integrated Daylighting Design

The Law Center accommodates varying program requirements within three interlocking L-shaped volumes (Fig. 5.2) organized around a daylit atrium space (Fig. 5.3) that serves as the primary means of circulation and aids in the passive ventilation of interior spaces (Fig. 5.4). For each of the three primary program types (office/classroom, library, and atrium/circulation), a unique high-performance facade was designed to meet interior daylighting objectives while controlling solar loads. The building can be cooled and heated in the extreme seasons with a radiant system, which maintains comfort more efficiently than conventional all-air

Fig. 5.1 Exterior view showing three facade types (office/classroom, library, and atrium). *Image credit* David Matthiessen, David Matthiessen Photography

Table 5.1 John and Frances Angelos Law Center

Owner	University of Baltimore
Completion date	April, 2013
Project type	Mixed-use Education Building (classroom, office and administrative spaces)
Gross floor area	17,837 m^2
Architect	Behnisch Architekten, Boston, MA, and Ayers Saint Gross, Baltimore, MD
Climate engineer	Transsolar, New York, NY
Lighting designer	MCLA Lighting Design, Washington, DC
Location	1401 N. Charles Street Baltimore Maryland 21201 United States

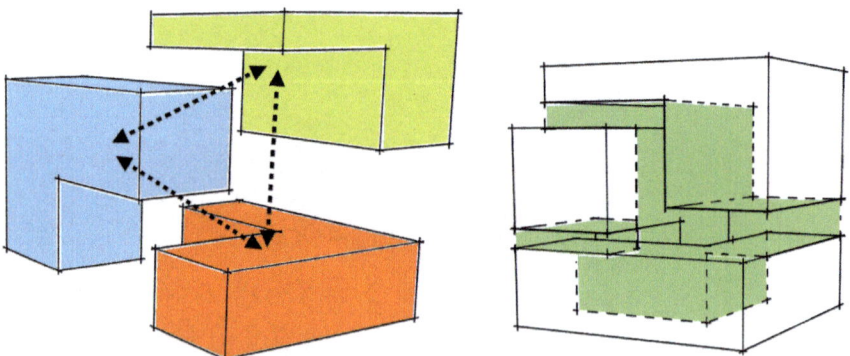

Fig. 5.2 Subdivision of the program into individual volumes, which interlock with a multi-story daylit atrium. The void space created by the separation of program volumes combined with the fragmentation of the building form into smaller pieces effectively reduces the scale of the project and increases the available surface area for fenestration. *Image credit* Behnisch Architekten

systems, and passively heated and cooled during more moderate seasonal conditions. Through the integration of automated exterior facade solar shading, interior daylighting, LED lighting, radiant space conditioning and mixed-mode ventilation, the building is predicted to achieve a 43% energy cost savings over an ASHRAE 90.1-2004 baseline building.

Fig. 5.3 Building section showing interior daylit atrium. Electrical lighting within 20 ft of facade glazing is controlled by photosensors to reduce output in response to available daylight. *Image credit* Behnisch Architekten

Cooling in the individual classrooms, offices, and most other occupied spaces is provided by an overhead thermally-activated concrete slab, reducing peak cooling loads and shifting a portion of these loads to nighttime. During the shoulder seasons, operable windows can be automatically controlled to allow outside air directly

Fig. 5.4 Building section showing daylit atrium space that serves as the primary means of circulation and aids in the passive ventilation of interior spaces. *Image credit* Behnisch Architekten

into the occupied zones. During these times, occupants are notified of via a green indicator light that it is possible to open a window. When outdoor conditions are not suitable for natural ventilation, the building management system prevents windows from opening (as a precaution to prevent moisture in humid air from condensing on the concrete slab) and a Dedicated Outside Air System (DOAS) with enthalpy wheel heat recovery is used to deliver minimum outside air to the occupied spaces to maintain air quality and provide dehumidification.

5.2.2 Office/Classroom Facade

The building exterior is clad with three distinct facade types: the office/classroom facade, the library facade, and the atrium facade. Each is discussed in detail in the following sections. The office/classroom facade is a glazed aluminum unitized curtain wall, with alternating punched window openings and solid aluminum panel units (Fig. 5.5).

Fig. 5.5 Office/classroom facade showing frameless glass screen wall designed to shield exterior automated venetian blinds. *Image credit* Behnisch Architekten

Punched windows include sections of operable window, ensuring all office and classroom spaces have access to natural ventilation. All glazed openings are shaded on the exterior using automated venetian blinds (80 mm wide slats) (Fig. 5.6) that can be positioned to fully block solar penetration to the building

Fig. 5.6 Office/classroom facade (*exterior view*) showing zoned sections of automated venetian blinds. *Image credit* Behnisch Architekten

interior. Slats have micro-perforations (approximately 6% of surface area) to allow vision through the slat even in the closed position. When in "closed" position, the upper 1/3 of blinds are configured to be at a 31° tilt angle for daylight redirection, while the lower 2/3rd of the blinds are at a 60° tilt angle to reduce glare and block solar radiation. Protecting these exterior blinds is a frameless glass screen wall, supported by brackets from the facade. This glass rain screen protects the shading from high winds in the upper stories and serves to unify the reading of the primary volumes that constitute the building's formal approach. Figures 5.7, 5.8 and 5.9 show the office/classroom facade in natural ventilation, heating, and cooling mode respectively.

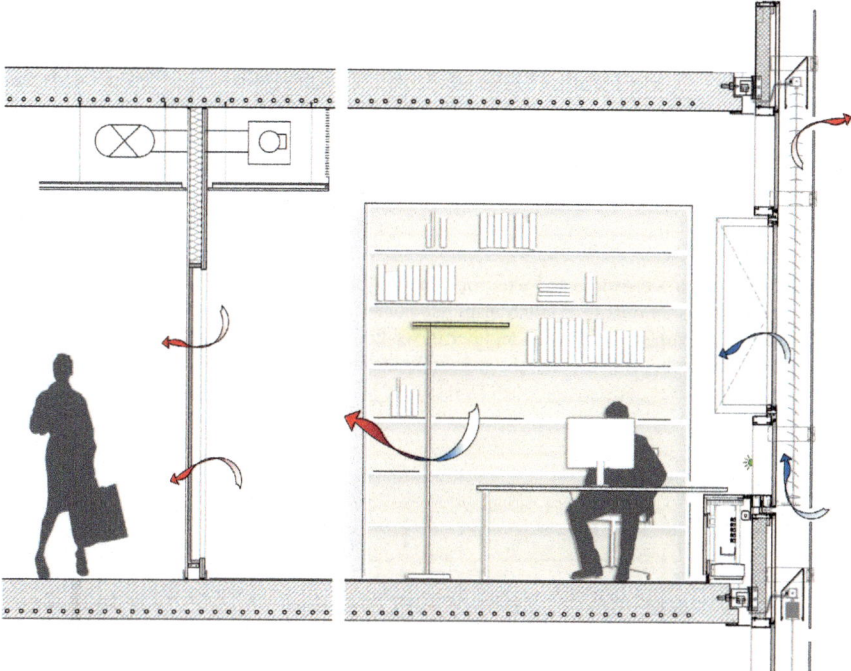

Fig. 5.7 Natural ventilation mode. Indicator lights signal the window may be opened for natural ventilation. Outdoor air enters the operable window through gaps in the glass screen, cools the room, and transfers to the atrium through custom air-transfer devices in the wall. *Image credit* Behnisch Architekten

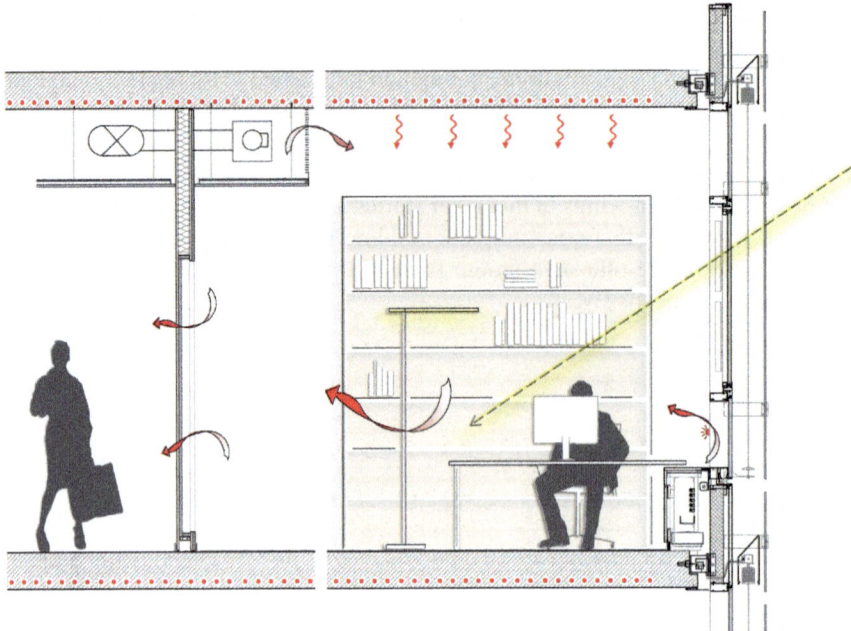

Fig. 5.8 Heating mode. Automated exterior venetian blinds are retracted. Minimal mechanical ventilation maintains air quality, while transmitted solar radiation provides base load space heating supplemented by overhead active slab and perimeter finned tube convectors. Offices are lit solely with unique stand-alone LED task lamps with daylight sensors. *Image credit* Behnisch Architekten

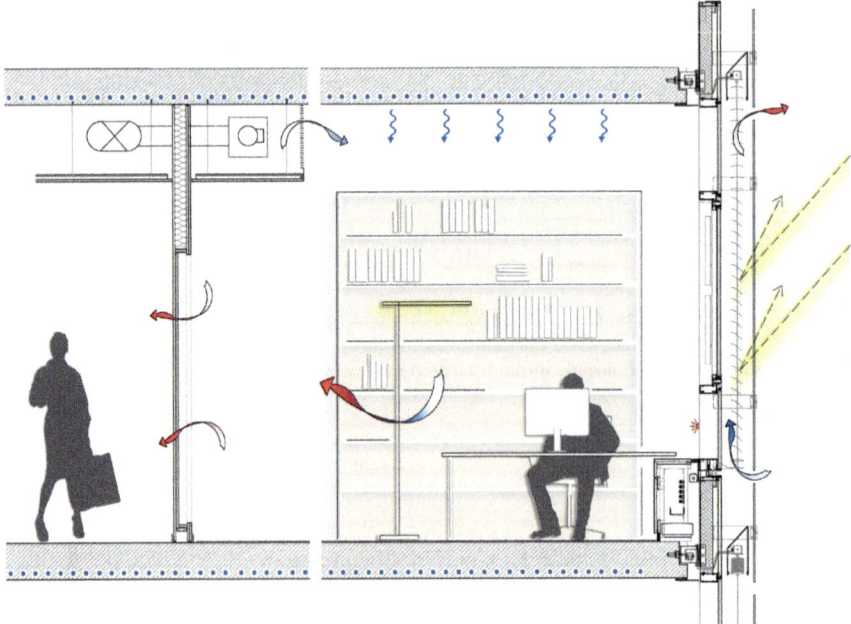

Fig. 5.9 Cooling mode. Automated exterior venetian blinds are deployed to reject solar heat gains into the facade cavity. Minimal mechanical ventilation maintains air quality and humidity, while overhead active slab provides space cooling. *Image credit* Behnisch Architekten

5.2.3   Library Facade

The second facade type is the library facade, also a glazed aluminum unitized curtain wall. However, in this case all of the units are glass treated with varying types of ceramic frit. Over the library facade the frit covers approximately seventy percent of the wall, protecting the interior from solar gain. One-half of the panels are fully fritted, and the other half are coated with a custom gradient frit pattern that alternates a half-floor height every other panel, creating a three-dimensional 'woven' effect (Fig. 5.10). The frit gradient dissipates and becomes transparent at the top of the glazing, with the objective of maintaining high levels of daylight transmission from the upper units and unoccluded views to the outside in the lower

Fig. 5.10 Library facade showing three-dimensional 'woven' effect created by an alternating half-floor height gradient frit pattern. *Image credit* David Matthiessen, David Matthiessen Photography

Fig. 5.11 The alternating pattern of window penetrations is integrated with the internal programming of the library perimeter zone. *Image credit* Behnisch Architekten

units (Fig. 5.11). Alternating panels have operable awning windows to enable natural ventilation in the library spaces.

5.2.4 Atrium Facade

The third facade type, the atrium facade, is an all-glass multistory curtain wall supported on a steel frame that spans between the building volumes (Fig. 5.12). Automated operable windows at each floor level introduce natural ventilation into the atrium and serve as make-up air inlets for the emergency smoke exhaust system. External fixed louvers on the south and west exposures protect the atrium from direct solar gains while allowing directs views to the exterior.

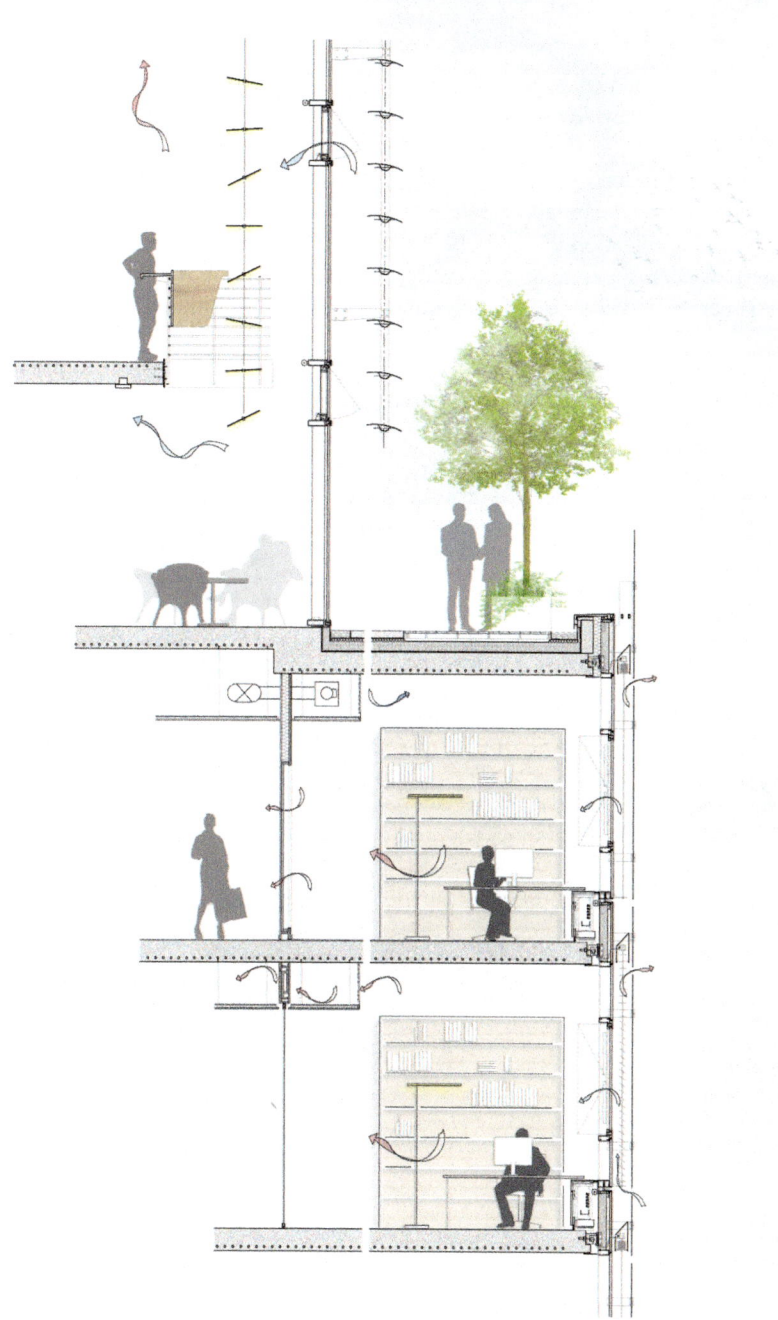

Fig. 5.12 Atrium facade wall section. *Image credit* Behnisch Architekten

5.3 NewActon Nishi

NewActon Nishi is a mixed-use commercial and residential development located within Canberra's NewActon precinct. The commercial office building is a 10-story tower attached to a residential housing block (Fig. 5.13). From early stage design, the project was guided by a client objective of achieving a 6 star Green Star design rating. The design team addressed this general goal for high performance by exploring the application of a number of passive design strategies integrated with low-energy systems and on-site renewable energy generation. Excluding renewable energy generated on site, in 2015 the building resulted in a measured (and publicly

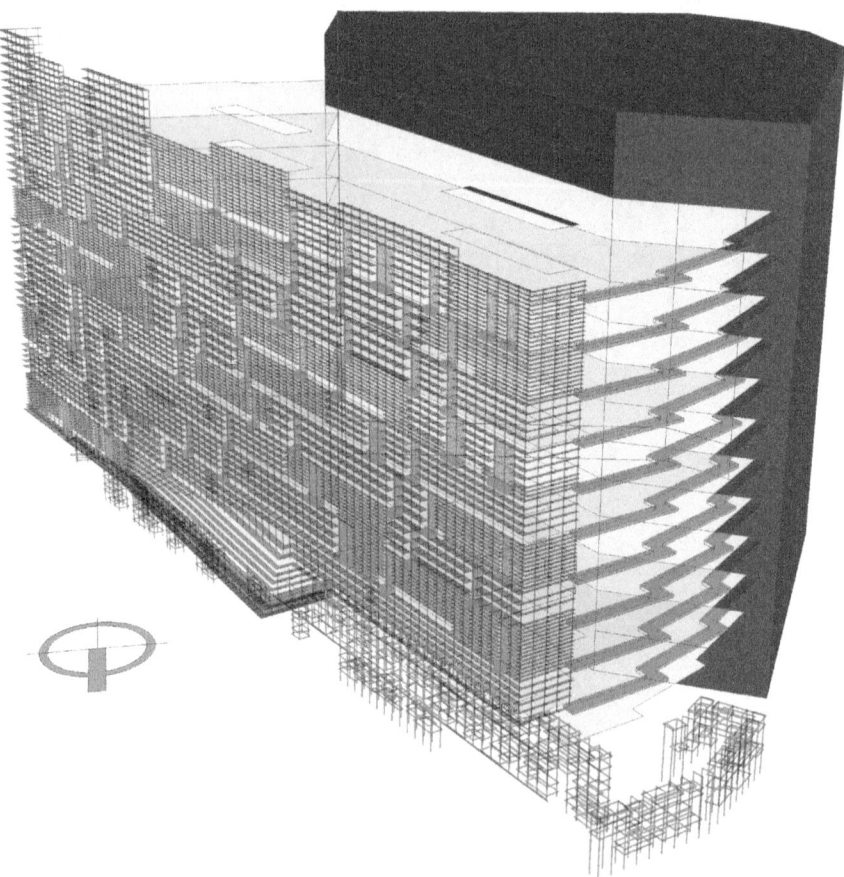

Fig. 5.13 Building massing. The office tower section of the mixed-use project is attached to a multi-family residential block (*dark grey*)

Table 5.2 NewActon Nishi

Owner	The Molonglo Group
Completion date	2013
Project type	Commercial office building
Gross floor area	22,500 m^2
Architect	Fender Katsalidis
Key collaborators	Arup, AWT Consulting Engineers, Oculus, PBS Building Group
Location	Canberra, ACT

disclosed via the Australian Government's Commercial Building Disclosure (CBD) program) annual energy consumption of 1,661,000 kWh (74 kWh/m^2-year) and an annual carbon emission intensity of 46 kgCO^2-e/m^2, resulting in a 5 star NABERS energy rating and making it one of the most resource efficient commercial buildings in Australia (Table 5.2).

5.3.1 Integrated Daylighting Design

The floor plate of the 10-story office tower is sidelit on three sides by a floor-to-ceiling glazed facade curtainwall [spectrally-selective low-e facade glazing (VLT 62%, SHGC 0.28, u-value 1.64 W/m^2 K)]. Facade glazing is shaded by external fixed horizontal wood louver screen with a north-east facing solar orientation (Fig. 5.14), and engineered to provide sufficient solar control to enable the application of passive and low-energy cooling strategies as an alternative to forced-air HVAC (Figs. 5.15 and 5.16). The blocking angle for the louvers was calculated to limit peak solar gain to 60 W/m^2. This was done so that a high efficiency/low temperature under-floor system could be utilized while still meeting peak cooling demand. Analysis was done in proprietary software to study various external shading strategies (horizontal and vertical) and glazing combinations using Canberra climate data. A cleaning access way is integrated into the screen (Fig. 5.17) to enable periodic maintenance and an irrigation system waters built-in planters using water captured from the roof. Additional glare control is provided by manually operated interior roller shades (VLT 6–9%).

Daylighting studies were performed to confirm that a Daylight Factor (DF) of 2% or higher was achieved for a minimum of 30% of the net leasable floor area, a requirement of the Green Star rating system Indoor Environmental Quality (IEQ) daylight credit, which awards points ranging from 1 to 3 based on minimum percentages achieved of 30, 60 and 90% respectively. Interior electrical lighting uses energy efficient LED fixtures and high efficiency T5 fixtures with electronic ballasts and achieves a lighting power density of less than 7 W/m^2.

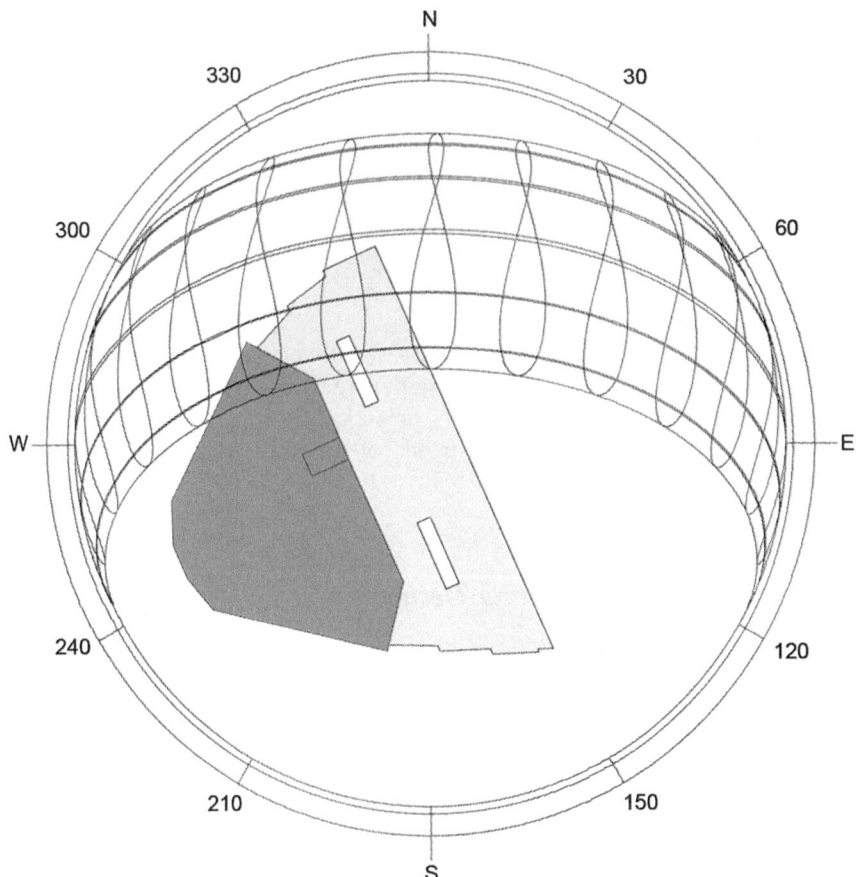

Fig. 5.14 Plan view of office floor plate with annual sun path overlay. Note site location is south of the equator (latitude = −35.3, longitude = 149.1). The office tower has two 10-storey atriums that act as the return-air shafts for the air conditioning, as well as bringing daylight deeper into the floor plate for the upper floors

Canberra's diurnal temperature range of 20 °C throughout the year is ideal for mixed-mode ventilation utilizing night purge to cool the exposed concrete thermal mass of the building at night. Operable windows are integrated into the facade (Fig. 5.17) and are automatically controlled by the building management system (which is integrated with an on-site weather station that monitors outside air temperature, humidity, wind speed and direction) to open whenever the outside air conditions are favorable. Occupants have the ability to manually override window operation for mixed mode ventilation in the perimeter zones if desired with a switch control panel on the facade, which controls a bank of 3–4 windows serving a 15 m^2

Fig. 5.15 Exterior view of the exterior wood solar control screen. The wood (Blackbutt or *Eucalyptus pilularis*) is considered a sustainable building material due to is wide availability and quick growth, and is designed to weather in place naturally. *Image credit* Carl Drury Photography

perimeter zone. Conditioned air is delivered by an under-floor air displacement system, which supplies air at low velocities evenly across the floor plate through circular grille diffusers in the floor that allow occupants to increase or decrease the air flow in their desk area. The ventilation scheme is assisted by two 10-storey atriums atria, which act as return airshafts and provide additional daylighting and views to the sky to the core zones of each floor plate.

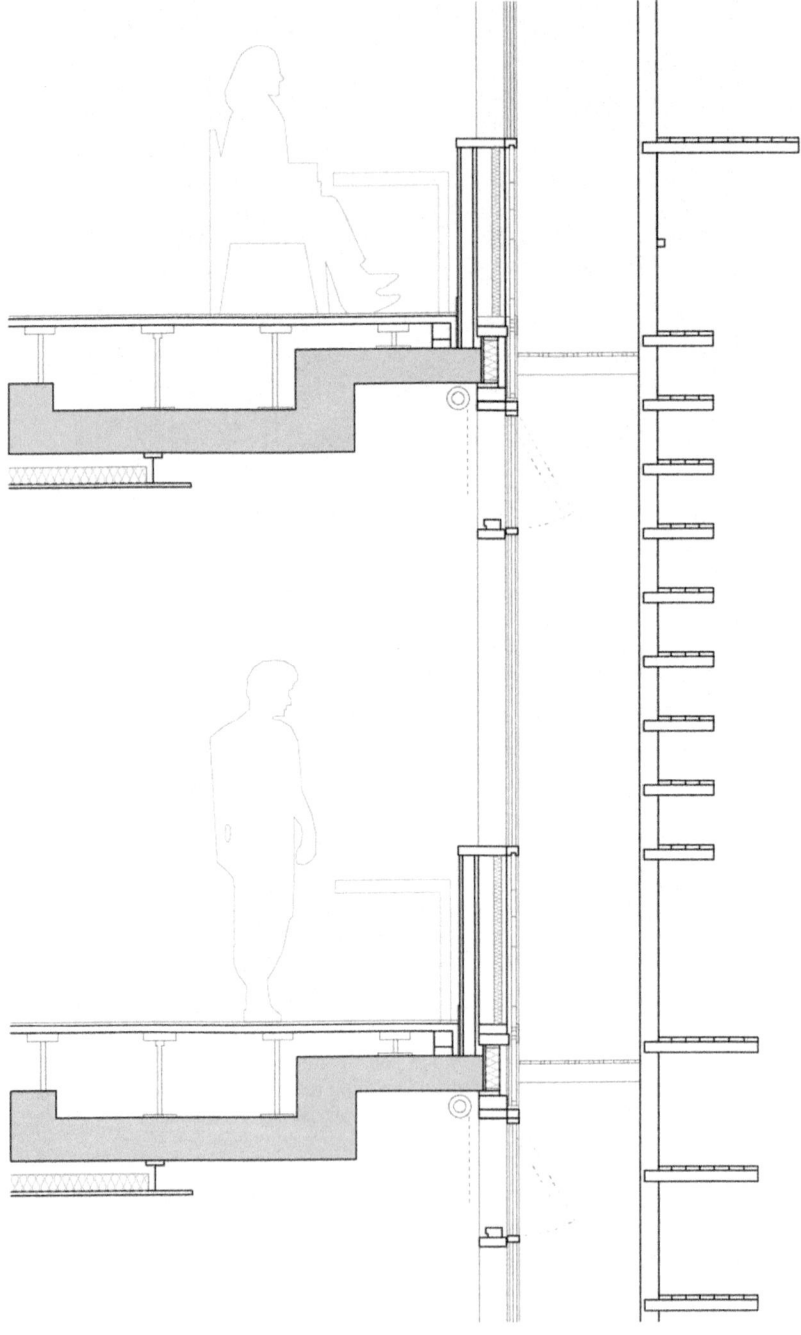

◀**Fig. 5.16** Wall section cut through building facade and perimeter zone. Exposed concrete ceilings provide thermal mass to enable effective thermal storage for night flush ventilative cooling. The blocking angle for the exterior fixed louver screen was calculated to limit peak solar gain to 60 W/m^2. Image drawn by Sue Long Lee

Fig. 5.17 View perpendicular to facade glazing showing external facade shading screen and operable transom windows used for automated natural ventilation. *Image credit* Carl Drury Photography

5.4 NREL Research Support Facility (RSF)

The Research Support Facility (RSF) (Fig. 5.18) serves as a case study in the application of integrated daylighting and energy simulation to design a large, Zero Net Energy (ZNE) 20,439 m^2 (220,000 ft^2) office building. Located on the National Renewable Energy Laboratory (NREL) campus in Golden Colorado, the RSF was envisioned "...to be a showcase of sustainable high-performance design to demonstrate the integration of high performance building design and practices in a replicable manner, showcase technology advances, and capture the public's imagination for renewable and energy efficient technologies." (Pless and Torcellini 2011). An Energy Use Intensity (EUI) target was set at 110 kWh/m^2 (35 kBtu/ft^2/ year), 50% below ASHRAE 90.1-2004, for a building that accommodates 820 people and includes a data center that can serve the entire NREL campus. The EUI and LEED (v2.2) Platinum performance goals were written into the solicitation process (NREL 2012) and contractually required in the project delivery model. The outcome was a performance-based design build process which led the design team

Fig. 5.18 The National Renewable Laboratory (NREL) Research Support Facility (RSF), south facade. Exterior vertical metal panels attached to opaque sections of the facade operate as a transpired solar collector to passively pre-heat ventilation air during the heating season. This image has been reprinted with permission from National Renewable Energy Laboratory

Table 5.3 National Renewable Energy Laboratory Research Support Facility (RSF)

Owner	U.S. Department of Energy National Renewable Energy Laboratory
Completion date	June, 2010
Project type	Large office building
Gross floor area	20,600 m^2
Architect	RNL, Denver, Colorado
Key collaborators	Stantec Consulting, RNL lighting design
Location	National Renewable Energy Laboratory (NREL) campus in Golden, Colorado

to integrate simulation-based feedback into early stage design through construction, as well as to develop a congruent process for detailed Post Occupancy Evaluation (POE) to compare performance in use with design intent (Table 5.3).

5.4.1 Integrated Daylighting Design

The objective of providing sufficient daylight and views to all 820 occupants led the design team to develop a prototype large office building that represents a significant departure from the conventional U.S. practice of a deep floor plate, air-conditioned building with a sealed and highly-glazed building envelope. Distribution of the program across two relatively narrow floor plates 18.3 m (60 ft) with open-office

Fig. 5.19 Third floor interior. A narrow floor plate [18.3 m (60 ft)] and an open-office workstation layout enable the potential for daylighting and natural ventilation for all occupants. This image has been reprinted with permission from National Renewable Energy Laboratory

workstation layouts enables the potential for daylighting and natural ventilation for all occupants (Fig. 5.19). The building is oriented to minimize solar exposure from east and west and maximize south and north-facing facade area for passive solar heating and daylighting. Window apertures (Fig. 5.20) are sized in response to solar orientation to provide daylighting while minimizing unwanted heat losses/gains and are individually subdivided to address multiple functions. These functions include exterior solar shading to provide solar/glare control in summer and permit direct gain solar heating in winter, daylight redirection for core-zone lighting via static, optical light-redirecting louvers (Fig. 5.21), provision of views to the exterior, and operable window zones for both automated and occupant-controlled natural ventilation. Windows on the south facade are subdivided into a lower view zone (SHGC = 0.23, VLT = 0.43) and upper daylight zone (SHGC = 0.38, VLT = 70). The Window-to-Wall ratios of the various facades are: (south = 0.30, east = 0.32, west = 0.31, north = 0.21), which represent a more thermally efficient alternative to fully glazed facades typical of many daylit buildings (see Chap. 1, Figs. 1.10 and 1.11).

The RSF also represents an early example of use of integrated daylighting and energy simulations to inform design (Guglielmetti et al. 2010, 2011). Radiance simulations using an iterative approach applied to a typical office floor were performed to help refine parameters including floor plate depth, window-to-wall ratio, window head height, subdivision of the window into daylight and view zones, and glazing visible light transmittance. The daylighting strategy is closely integrated

Fig. 5.20 The windows on the south facade are divided into an upper "daylight zone" and lower "view zone." Exterior shading is designed to provide solar and glare control to the view zone to avoid the need for interior shading devices. This image has been reprinted with permission from National Renewable Energy Laboratory

with automatic, continuously dimming, daylighting controls in all daylit zones as well as occupancy controls and high-efficiency electrical ambient and task lighting. Simulation also played a key role in informing the zoning of electrical lighting fixtures, how they should be controlled (e.g. switching vs. automated daylight-dimming) and to predict the potential for annual energy reduction. To visualize the annual daylight distribution in the space, floor plates were rendered in falsecolor showing the outcome of a radmap-generated "cumulative sky" in Radiance. Radmap is a Radiance-based tool used to generate annual irradiance and illuminance maps (Anselmo and Lauritano 2003). Figure 5.22 presents an example of the cumulative annual sky falsecolor rendering. The process is described in detail in (Guglielmetti et al. 2010). For the open offices, 269 lx (25 footcandles) and a 4:1 maximum-to-minimum illuminance ratio were selected as the ambient workplane illuminance and uniformity criteria and an additional 215–323 lx (20–30 fc) was specified for task lighting to meet the IESNA office recommendation of 323–538 lx (30–50 fc) for general office task lighting overall. The total predicted lighting EUI resulted in 7% of the final as-built energy model total EUI of 105 kWh/m^2-year (33.3 kBtu/ft^2-year).

The climate of Golden Colorado has cold winters, hot summers and receives, on average, about 240 clear sunny days per year. Through effective solar control and a high-efficiency building envelope, the RSF is capable of meeting heating and

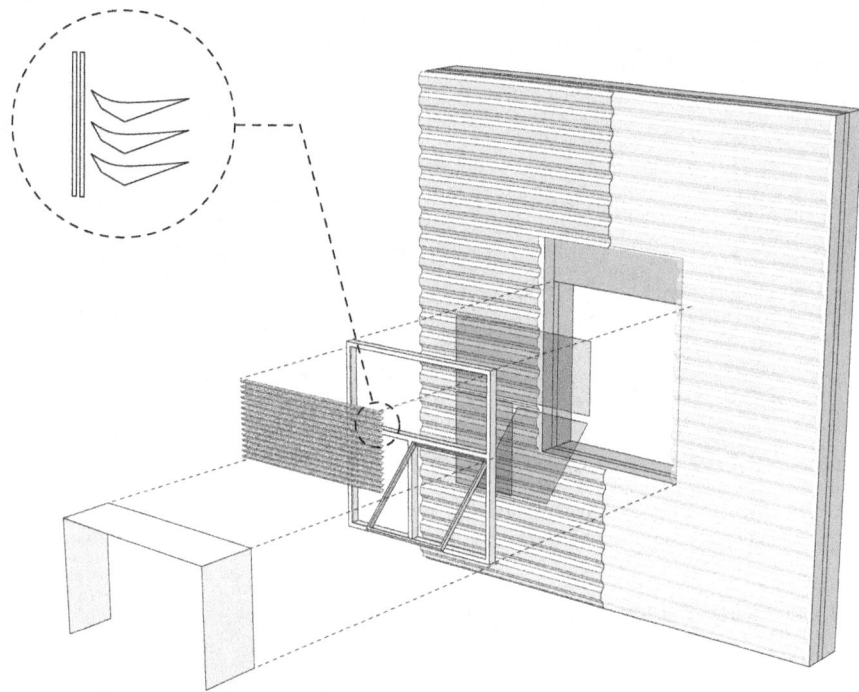

Fig. 5.21 The daylight zone has a high VLT glazing (VLT = 0.70) and integrated static optical lighting redirecting lover system (*circled*, see Chap. 3, Sect. 3.2 for a detailed description) to redirect direct beam sunlight towards the ceiling to increase daylight penetration. Occupants can open windows for ventilation and cooling during appropriate seasonal conditions. View zone windows are triple-glazed and have improved thermal breaks to improve thermal performance and occupant thermal comfort. Daylight zone windows are double-glazed. Image drawn by Sue Long Lee

cooling needs while avoiding a forced air heating and air conditioning system. Management of occupant thermal comfort is addressed through the integration of exposed thermal mass, radiant heating and cooling, natural ventilation (with automated night-purge) and direct and transpired passive solar heating. A demand controlled Dedicated Outside Air System (DOAS) provides fresh air from a raised floor when windows are closed on the hottest and coolest days. Ventilation air is distributed through an under-floor air distribution system. During the heating season, outside ventilation air is passively preheated via a transpired solar collector on the south facade integrated with a thermal labyrinth in the concrete foundations of the office wings (see Sect. 5.4.2). Evaporative cooling and energy recovery systems further reduce outdoor air heating and cooling loads.

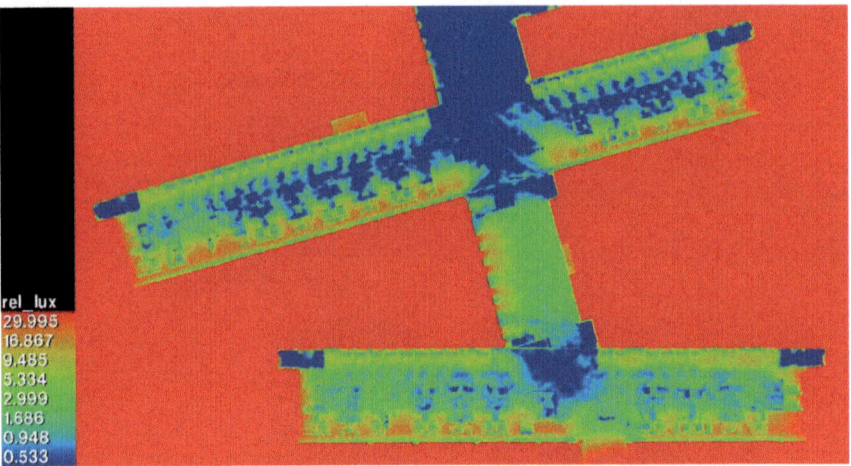

Fig. 5.22 NREL RSF cumulative annual sky falsecolor rendering. *Image credit* R. Guglielmetti/NREL. Reprinted with permission of the National Renewable Energy Laboratory

5.4.2 Solar Thermal Energy Harvesting

In addition to energy harvesting using solar photovoltaic systems, solar thermal energy systems can be applied to opaque areas of the facade to provide thermal energy, which is typically applied to support space heating or hot water production. Figures 5.23 and 5.24 show the application of a transpired solar collector on the south facade of the NREL RSF building located in Golden Colorado. A transpired solar collector is a simple system that consists of a metal sheet perforated with small holes that is installed several inches from the facade to create an air cavity. The metal sheet is heated by direct solar radiation and ventilation fans are used to create a negative pressure in the cavity, drawing in the heated air through the small perforations (Fig. 5.25). The heated air can then be ducted inside the building, or in the case of the RSF, the heated air is directed into a thermal labyrinth (staggered concrete walls) in the foundation of the RSF (Fig. 5.24), where it is used to "charge" the thermal mass of labyrinth and pre-heat ventilation air during the heating season that is then drawn into the occupied areas of the building. The combined effect of the solar collector and labyrinth can warm outside air by 2.8–5.6 °C (5–10 °F) (NREL 2012).

Fig. 5.23 Transpired solar collector installed on south facade of the RSF. *Image credit* Dennis Schroeder, NREL

Fig. 5.24 Transpired solar collectors on south facade of building which supply passively-heated air to the thermal labyrinth below the NREL RSF building. *Image credit* RNL

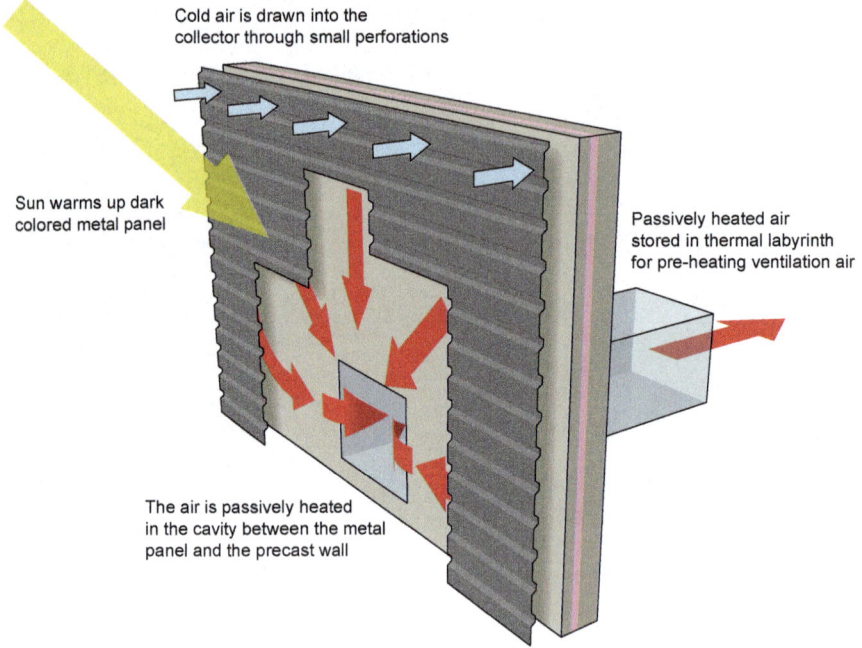

Cold air is drawn into the
collector through small perforations

Sun warms up dark
colored metal panel

Passively heated air
stored in thermal labyrinth
for pre-heating ventilation air

The air is passively heated
in the cavity between the metal
panel and the precast wall

Fig. 5.25 Diagram showing how the transpired solar collector works. *Image credit* RNL

5.4.3 Comparison of Design Intent with Performance in Use

In contrast to many projects that claim effective use of daylighting for energy
reduction, the delivery of the RSF included detailed commissioning to evaluate the
fidelity of daylight design intent and model predictions with measured energy data.
Figure 5.26 compares whole-building operational lighting power for a typical day
against the maximum lighting power allowed by the energy efficiency standard
ASHRAE 90.1, the maximum expected power for the system (installed lighting
load without daylight dimming), and the simulation-based prediction of lighting
power. The figure shows the significant impact of the daylight dimming lighting
controls, which exceeded predicted reductions and reduced lighting power by over
80% from the ASHRAE 90.1 baseline for energy efficient lighting during daylight
hours. The comparison between design assumptions and performance in use also
reveals a significant discrepancy in nighttime electrical lighting use, where nighttime lighting power was higher than predicted due to scheduled cleaning of the
building.

Figure 5.27 investigates the effort to integrate daylight responsive electrical
lighting control with the daylight availability achieved from the architectural daylighting design. The figure shows a single day with clear sky conditions. The
electrical lighting load is shown to decrease in proportion to the available daylight
as measured by global exterior horizontal illuminance.

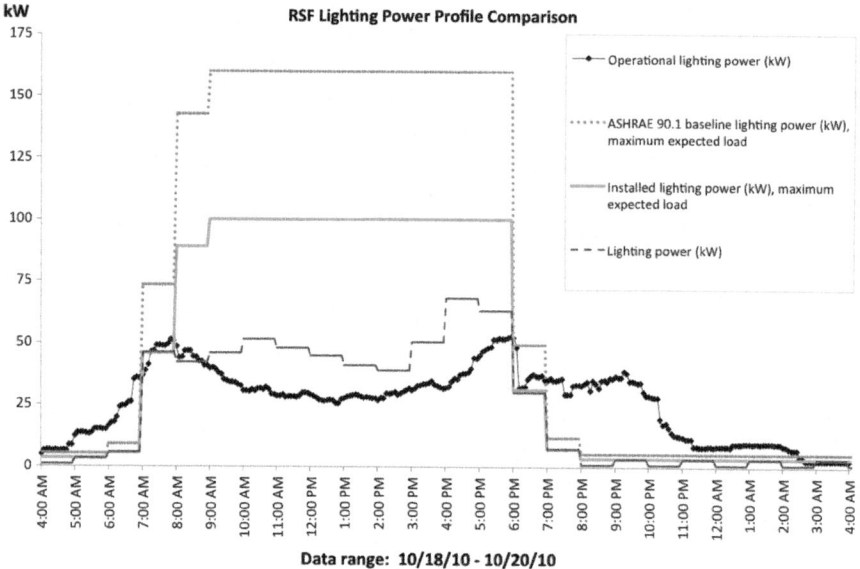

Fig. 5.26 RSF lighting power comparison. This image has been reprinted with permission from the National Renewable Energy Laboratory

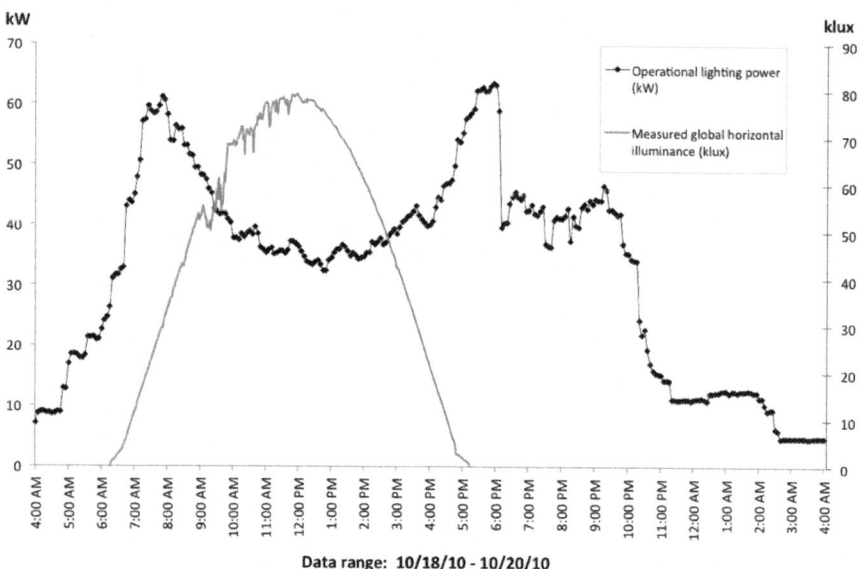

Fig. 5.27 RSF lighting power profile, plotted with global exterior horizontal illuminance. This image has been reprinted with permission from the National Renewable Energy Laboratory

5.5 Bullitt Center

The Bullitt Center (Figs. 5.28 and 5.29) serves as a case study in the integration of daylighting within a broad set of resource efficiency and indoor environmental quality design goals. Promoted by its developer as "the Greenest Office Building in the World," the project is a 6-story office building designed to comply with the Living Building Challenge rating system, which requires post-occupancy verification of ZNE performance in use, along with a range of other environmental performance goals including zero net water, zero net carbon, and the creation of a "beautiful" built environment. In comparison to a typical Seattle office building (EUI) of about 227 kWh/m^2-year (72 kBtu/ft^2-year), the Bullitt center is designed to achieve an EUI of 50.5 kWh/m^2-year (16 kBtu/ft^2-year), with the annual energy consumed by the building offset by electricity generated from a 242 kW rooftop solar photovoltaic (PV) array (Fig. 5.28). Due to the scale of the project relative to the site footprint, surface area for locating the PV array on the building was limited and in competition with surface area required for fenestration. Therefore, driven by the spatial constrains of the site, local climatic conditions, and the ZNE performance target, the design team worked to develop a highly efficient building envelop

Fig. 5.28 Exterior view of the Bullitt Center from street level showing exterior shading deployed on facade in direct sun and exterior shading retracted for facade in shade

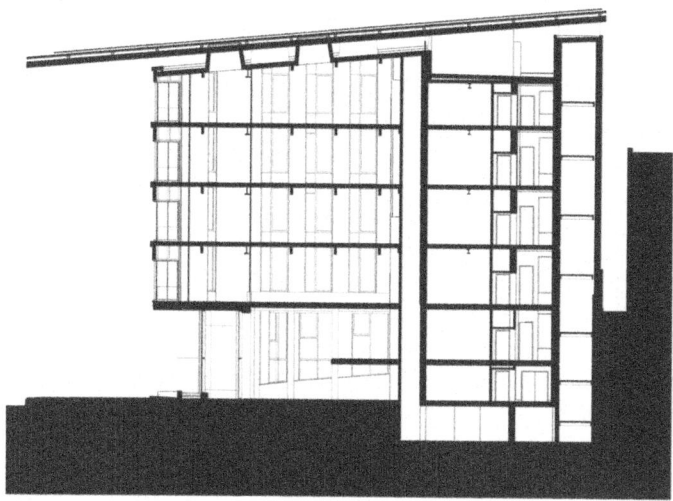

Fig. 5.29 Building section. *Image credit* The Miller Hull Partnership

Table 5.4 The Bullitt Center

Owner	The Bullitt Foundation
Completion date	April 22, 2013
Project type	Medium office building
Gross floor area	4645 m^2
Architect	The Miller Hull Partnership
Key collaborators	PAE, Point32, Schuchart, Foushee, Solar Design Associates, Northwest Wind and Solar, DCI Engineers, Luma Lighting Design, 2020 Engineering, Berger Partnership, RDH
Location	1501 East Madison Street Seattle, WA 98122 USA

to minimize loads and enable the application of passive environmental control strategies of daylighting, direct gain solar heating, natural ventilation, and night-flush cooling. These strategies are combined with low-energy mechanical systems (ground source heat pumps, in-floor radiant heating/cooling, and a Dedicated Outdoor Air System (DOAS) with heat recovery). During the first 12 months of occupancy (May 2013–April 2014), the building's measured EUI was 29.6 kWh/m^2-year (9.4 kBtu/ft^2-year) (Peña 2014). This outcome (41% lower than designed), was due in part to the relatively low occupancy of the building during the first year (about 50% occupied) (Peña 2014) (Table 5.4).

5.5.1 Integrated Daylighting Design

Daylighting design goals were central to the performance-based design process used to develop the final design scheme and impacted decisions at all levels, including building form and massing, floor-to-ceiling height, fenestration configuration, interior zoning and programing, and the configuration of structural elements. As the first step in performance-based design, performance goals were defined at the beginning of the design process. The Living Building Challenge requires that "Every occupiable interior space of the project must have operable windows that provide access to fresh air and daylight," and that workstations must be no more than 9.14 m (30 ft) from windows (ILFI 2012). These requirements led the design team to initially explore "O" and "U" shaped atrium schemes in order to "get more of the floor plate close to a source of daylight and fresh air and to drive daylight deep into the building's core" (Peña 2014). However, these schemes were abandoned in favor of a "T" shaped scheme following performance feedback from Radiance-based daylighting simulations showing that the atrium schemes resulted in minimal additional daylighting to lower floors and limited the roof area available for the PV system. The T shaped scheme (Fig. 5.30), has a 21 m (69 ft) wide floor plate with a 6.4 m (21 ft) wide central service core situated within two 7.3 m (24 ft) wide perimeter zones. The scheme is sufficient to achieve the distance-to-window requirement for work zones while also reducing envelope heat loss due to a significantly lower surface-to-volume ratio compared with either atrium scheme.

   Daylighting of the Bullitt center follows a conventional sidelighting pattern with floor-to-ceiling vertical windows spaced between sections of opaque wall (Fig. 5.31). Simulation studies using an overcast sky design condition where used

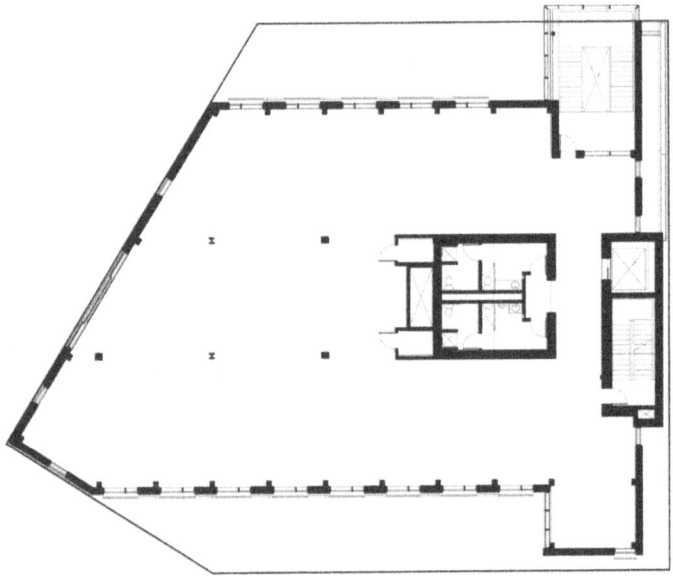

Fig. 5.30 Plan of 4th floor. *Image credit* The Miller Hull Partnership

Fig. 5.31 Sixth floor daylit perimeter zone. Note the absence of lighting fixtures on the ceiling. The installed electrical lighting power density is extremely low as a result of the decision to install minimal fixtures and require tenants to install supplemental electrical lighting if desired as a tenant improvement. On site observations (made by the author) revealed that no tenants have installed additional electrical lighting as of the publishing of this book. This is perhaps the most reliable indicator of a space achieving "daylight autonomy"

to evaluate the daylighting potential of a number of window configurations relative to a fully glazed facade with a 3.5 m (11.5-ft) floor-to-floor height using the Daylight Factor (2% threshold) as a performance indicator. Studies showed that the fully-glazed facade led to minimal improvement over a more energy-efficient configurations including opaque wall. However, an increase in floor-to-floor height to 4.22 m (13'-10", 13'-1" ceiling) were found to significantly improve the percentage of floor area that achieved 2% DF (from 23 to 62%). This simulation feedback, obtained in early stage design, led the team to petition the Department of Planning and Development for a departure from the local zoning code building height limit of 19.8 m (65-ft), to extend the floor-to-floor height of floors 3 through 6 (Fig. 5.29). The departure was granted through the Living Building Pilot, (City of Seattle 2016[1]) which is an ordinance put in place specifically for buildings attempting to meet the Living Building Challenge.

The building envelope was conceived as an "operable skin" to achieve daylighting goals while minimize thermal losses during winter and dynamically controlling solar heat gains in summer. In early stages or design, thermal energy simulations were used to examine design alternatives for both glazing and opaque wall thermal performance to reduce annual heating loads. Due to the significant impact of various glazing configurations determined through these studies, a

---

[1]http://www.seattle.gov/dpd/permits/greenbuildingincentives/livingbuildingpilot/.

Fig. 5.32 Exterior automated venetian blinds deployed to block direct sun during the cooling season. Note that the windows on the top floor do not require exterior shading due to the shading provided by the overhanging solar canopy

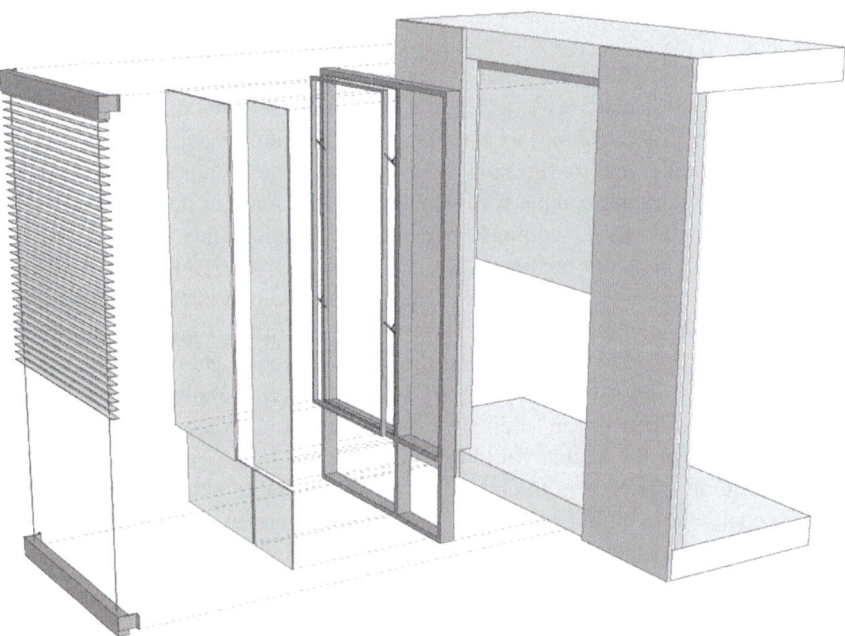

Fig. 5.33 Exploded view of facade aperture showing exterior automated venetian blinds, automated "pop-out" windows for natural ventilation, triple-glazed curtain-wall system (with 1 or 2 low-E coatings, argon gas fill, and warm edge spacers), and interior roller shade (partially deployed). Image drawn by Sue Long Lee

high-performance triple-glazed curtain-wall system was developed, resulting in an effective U-value of no more than 0.18 (Btu/h-ft^2 °F) and a SHGC of 0.59. The relatively high SHGC was possible due to the integration of automated exterior venetian blinds (Figs. 5.32 and 5.33), which enable the transmission of optimal levels of solar heat gains during the heating season and complete control of direct beam radiation during the cooling season.

5.6   New York Times Headquarters

The New York Times Headquarters (Fig. 5.34 and Table 5.5) is a 52-story, 139,355 gross m^2 (1.5 million gross square feet) commercial office building located in Manhattan, New York. Occupied in 2007 and extensively evaluated five years after occupancy, the project demonstrates significant electrical lighting energy savings relative to a similar code compliant building through the combination of design for daylighting, occupancy sensing and setpoint tuning. From early stage design, the owner and architectural design team collaborated with researchers at the Lawrence Berkeley National Laboratory (LBNL) to apply thoughtful architectural

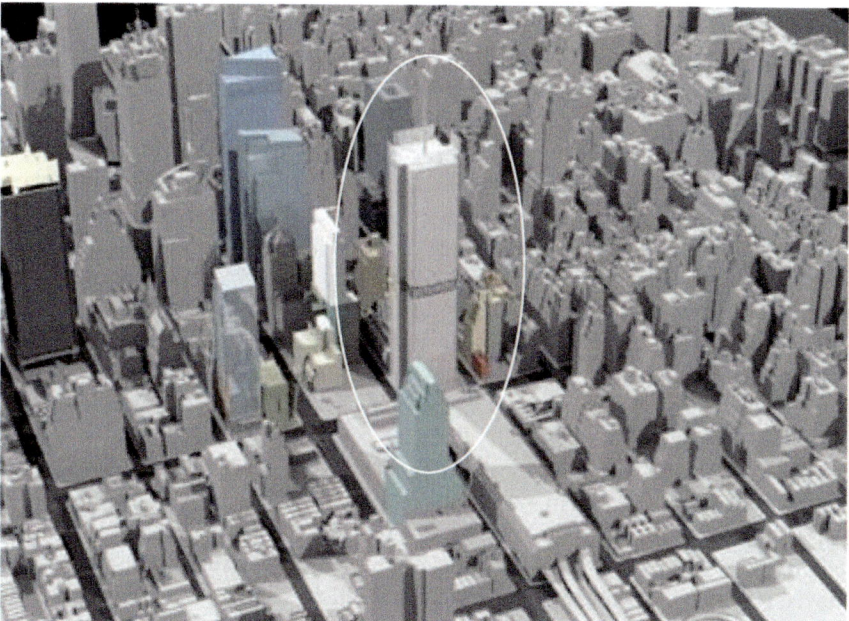

Fig. 5.34 Rendered view of the New York Times Building (*circled*) shown in the urban context in which it was modeled for daylighting and solar exposure analysis. *Image credit* LBNL

Table 5.5 The New York Times Headquarters

Owner	The New York Times Corporation, Forest City Ratner Companies
Completion date	July 2007, (POE completed in 2013)
Project type	Large office building
Gross floor area	139,340 m^2
Architect	Renzo Piano Building Workshop in association with FXFOWLE Architects
Key collaborators	Lawrence Berkeley National Laboratory, Loisos Ubbelohde, Gensler, Thorton Tomasetti, Flack and Kurtz, AMEC Construction Management Inc.
Location	1401 N. Charles Street Baltimore Maryland 21201 United States

design practices focused on daylighting and visual transparency with rigorous integration of automated shading technologies and controls in a collaborative process that occurred from early stage design through construction and commissioning and into operations.

Prior to construction, but after the primary glazing and fixed shading solutions had been developed, a 418 m^2 (4500 ft^2) full-scale fully furnished mockup was built and commissioned by the owner. With support from the NYSERDA and the US DOE the LBNL team then developed, evaluated and optimized automated interior shading and daylight-dimming lighting control technologies with the objective of optimizing the user's experience in the space, reducing installation challenges and performance uncertainty in the actual building. Novel, Radiance-based simulation techniques were used to develop granular, context-aware dynamic shading control algorithms, sensitive to sun position, an extensive grid of exterior illuminance sensors, and overshadowing effects of adjacent buildings. Due to the availability of the full scale mockup, and the size of equipment procurement for the project, the owners were able to engage with suppliers and challenged them to deliver unique new performance requirements for key light control and automated shading systems, which resulted in lower costs and increased market availability of these systems for more widespread implementation in future projects. LBNL undertook a year long field study after occupancy which compared measured data on one floor to simulated results from a similar, code-compliant building. The measured performance resulted in a 56% lighting energy savings (0.37 kWh/m^2-year (3.94 kWh/ft^2-year) across a 12.2 m (40-ft) deep perimeter zone), 24% total energy savings, and 21–24% reduction in summer peak demand (Lee et al. 2013). Beyond energy, occupant subjective survey data showed generally high levels of occupant satisfaction, comfort and acceptance with energy efficiency measures (Lee et al. 2013). The project is one of the most well-monitored and documented case studies on the benefits of designing buildings as integrated whole building systems (Lee et al. 2013).

5.6.1   Integrated Daylighting Design

The climate of New York City is classified as a humid continental climate,[2] with summer temperatures of 90 °F (32 °C) or higher recorded on average 18–25 days each summer. Even in climates with less extreme summers, control of excessive heat gains through windows governs facade design, and often leads to heavily coated glass, or smaller windows. Both options compromise daylighting potential, as well as the visual connection to the exterior for occupants and may still result in inefficient whole-building energy performance and the unnecessary oversizing of mechanical HVAC equipment. The design goal for the New York Times building was to create an optically transparent facade while still meeting whole-building energy efficiency goals.

To address the challenge of solar control, the design team developed a "second skin" of ceramic rods spaced 18 in. from the glazed facade and designed to block 50% of direct beam radiation (Fig. 5.35). Each rod is 1.52 m (5-ft) long and 41 mm (1–5/8-in.) in diameter. In addition to acting as an external shading device, the rods diffuse a portion of incident direct beam indoors. To improve the quality of occupant views, the vision portion of the facade (from 0.88 m (2.9-ft) to 2 m (6.6-ft) above the finished floor) is unshaded to allow occupants to have unobstructed views from seated and standing positions. Rods are spaced more densely above head height relative to below the view zone (Fig. 5.36). The exterior shading layer enabled the interior facade layer to be glazed from floor to ceiling (window-to-exterior-wall ratio of 0.76) with double-pane low-iron water-white high Visible Light Transmittance glazing (VLT = 0.75), which includes a spectrally

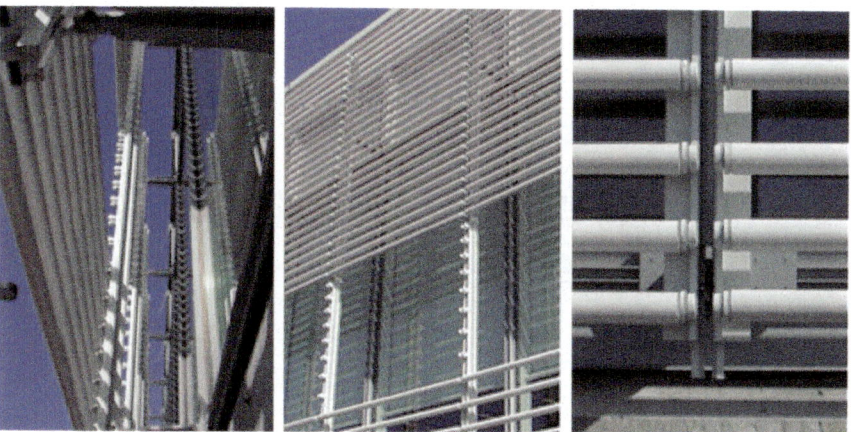

Fig. 5.35 Exterior facade of daylighting mockup. Photographs by: Voropat Inkarojrit. *Image credit* LBNL

---

[2]Köppen climate classification.

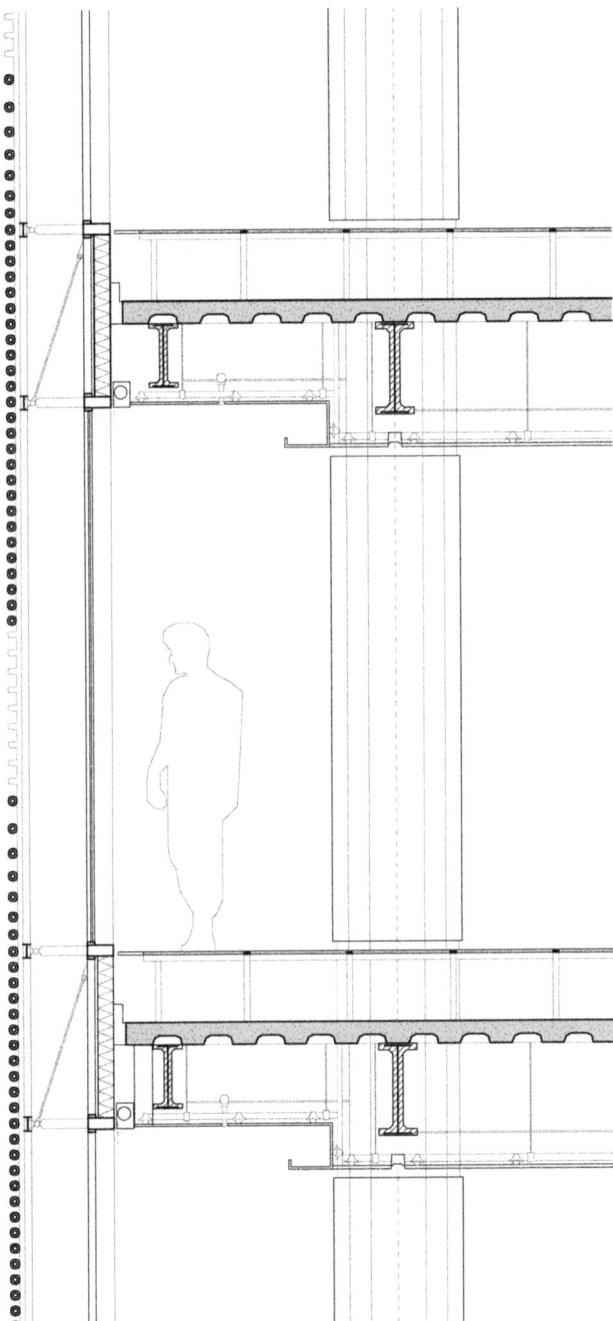

Fig. 5.36 Wall section. The ceiling height of 2.93 m (9.6 ft) is greater than typical U.S. commercial construction (2.74 m) and the ceiling height increases near the facade to 3.14 m (10.3 ft) to increase daylight penetration. Image drawn by Yang Li after wall section by Renzo Piano Architects

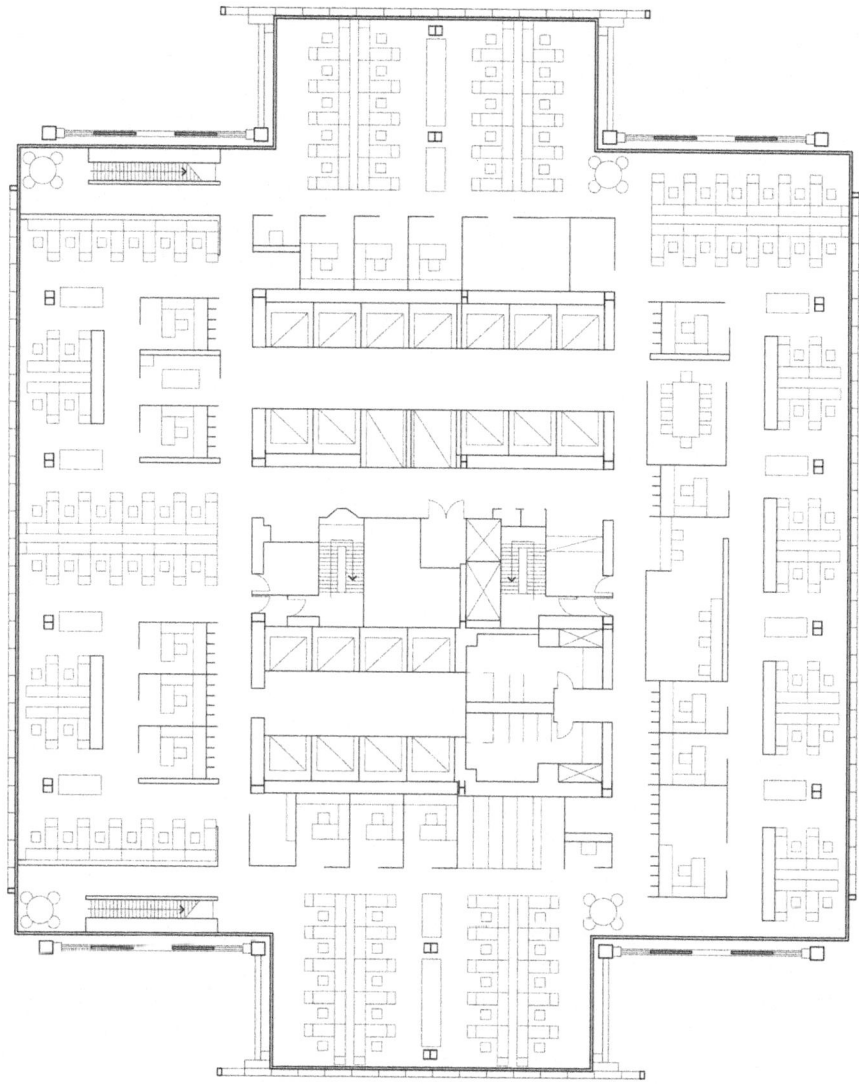

Fig. 5.37 Floor plan of typical office floor. Image drawn by Yang Li after floor plan by Renzo Piano Architects

selective low-e coating to further reduce solar heat gains. Limiting solar gains to the perimeter zones enabled the design team to specify an Underfloor Air Distribution System (UFAD) for space conditioning as a low-energy alternative to a conventional forced air HVAC system.

The floor plan layout is designed to preserve views and access to daylight for all occupants. Conference rooms and glazed private offices are located 7.6 m (25-ft) from the facade, leaving the perimeter zones available for open-office workstations

Fig. 5.38 Images of roller shade operation on February 23, 2004 in the daylighting mockup. *Left* view of southwest corner looking west; *middle* view of southwest corner looking southwest; *right* view of west facade near center of the mockup dividing the two areas. *Image credit* LBNL

with low partition heights [1.22 m (4 ft)], allowing views to the exterior as well as internally. The cruciform shape of the floor plate enhances daylight penetration and enables nearly all occupants to have views in three directions, preserving access to an unshaded view while shades are deployed on other facades to control glare and direct sun (Fig. 5.37).

While the exterior screen was considered effective for control of solar heat gains, glare control and daylight management were addressed through the application of a system of automated interior fabric roller shades (Fig. 5.38), and responsive electric lighting controls. While automated shades have been available for decades, intelligent and effective shade controls, which achieve measured cooling savings and electrical lighting energy savings and lead to satisfied occupants, have remained elusive. The challenge to effective controls lies in the need for reliable and granular control of shades to detect and respond to dynamic local glare conditions in real-time without sacrificing daylighting potential or occupant visual comfort. There is also a balance of minimizing glare with a low transmittance shade fabric versus daylight admittance that argues for higher light transmittance. The control operations should vary with orientation and height in a dense urban environment with adjacent buildings. The LBNL research team addressed this challenge by working with the owner, contractors and equipment manufactures to develop a system programmed to use sun position and real-time illuminance data from a network of sensors arrayed across the building skin to adjust shade positions to a range of heights determined to limit direct sun penetration to a specified distance from the perimeter glazing. Rather than simply deploying or retracting completely, the shades can be controlled to five present heights that align with vertical registration points on the facade.

The ambient electrical lighting system is integrated with the daylighting strategy and controlled on the assumption that occupants prefer daylight (when available), over electrical light sources. The installed lighting system consists of over 18,000 daylight-dimming fixtures, each of which is individually addressable from a software management system via a Digital Addressable Lighting Interface (DALI). The dimmable lights can be tuned based on personal preference on a granular basis to deliver light levels below the maximum output independent of daylight, In addition to daylight harvesting, dynamic control of shades and lighting also serves to reduce

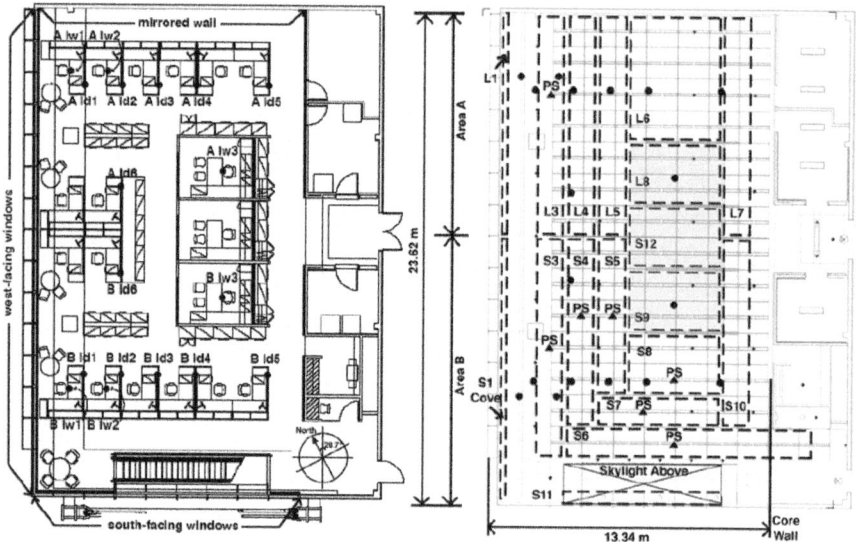

Fig. 5.39 Floor plan of full-scale mockup showing the location of the interior illuminance sensors (*left*) and reflected ceiling plan (*right*) showing the lighting zones and location of photosensors (PS *triangle symbols*). *Image credit* LBNL

loads on HVAC and can be used to participate in Demand Response (DR) activities, such as peak load reduction.

Prior to construction and product specification, a 418 m² (4500 ft²) full-scale fully furnished mockup (Fig. 5.39) was commissioned to develop, evaluate and optimize automated interior shading and daylight-dimming lighting control technologies with the objective of reducing installation challenges and performance uncertainty in the actual building.

> We used the mockup to develop our thinking and to evaluate a few examples of automated facade management and daylight harvesting systems.—Glenn Hughes, Director of Construction, The New York Times Company

5.6.2 Post Occupancy Evaluation

In 2011–2012, The New York Times Company collaborated with LBNL and the U. C. Berkeley Center for the Built Environment (CBE) to monitor the performance of the dimmable lighting, automated interior roller shades, and underfloor air distribution system (UFAD) as well as to conduct an Indoor Environmental Quality (IEQ) survey of building occupants. Electrical lighting energy was monitored directly from each lighting circuit. Figure 5.40 presents and example of how lighting energy use savings are attributed to each control strategy over a 24-h day.

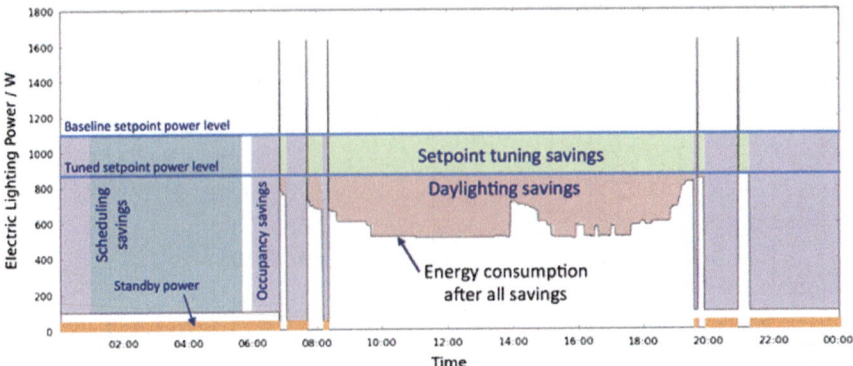

Fig. 5.40 Example of how lighting energy use savings are attributed to each control strategy over a 24 h day. Control strategies include scheduling, occupancy, setpoint tuning, and daylighting. *Image credit* LBNL

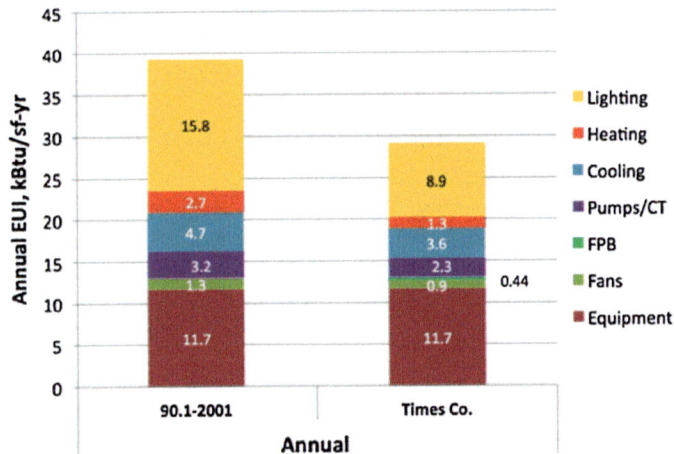

Fig. 5.41 Annual end use energy comparison; baseline overhead versus calibrated Times Building model. *Image credit* LBNL

A calibrated EnergyPlus model informed by operations data and measured data was used to determine HVAC energy use. Compared to a similar, code-compliant building, the measured performance resulted in a 56% lighting energy savings (0.37 kWh/m²-year (3.94 kWh/ft²-year) across a 12.2 m (40-ft) deep perimeter zone), 24% total energy savings, and 21–24% reduction in summer peak demand (Fig. 5.41) (Lee et al. 2013).

5.6.3 Lessons Learned

The LBNL post occupancy study not only collected performance data on the building and its occupants, but provided some key data from the field to assess progress towards more widespread use of daylighting systems that integrate smart lighting controls with automated shading. The project was widely discussed in the industry and resulted in two manufacturers offering new products i.e. DALI-based dimmable, addressable fluorescent lighting with daylight controls and automated motorized shading with facade specific shade controls. The team expected that the project highlights, the availability of extensive descriptive materials and the new products might result in a major increase in the use of these systems in highly glazed buildings. In general this has not been the case. Further review of lessons learned and "progress" in the 13-year period since the project was initially launched provides the following lessons and insights into gaps that remain.

1. Successful design and deployment of these systems requires a committed owner, a skilled design team, knowledgeable contractors and informed occupants, as well as the appropriate products and systems. Lacking any one of these over the multiyear period of design, construction and occupancy can create significant challenges to successful implementation.
2. While the technologies have evolved since this project, and in some cases provide new more effective and lower cost options, particularly in the LED lighting/controls field, the technology specification process remains complex. That function in the Times project was facilitated by LBNL's support and the work in the mockup; it is not yet easily assimilated into standard tools and design practice. In the short term, too many incompatible technical options for sensors, dimming controls, shading fabrics, etc. confuse design professionals and owners. Interoperability, better design tools, new standards and greater training, experience and expertise on the part of contractors should eventually reduce the scale of this problem. The "wild west" of the Internet-of-Things offers great promise longer term but presents a steep learning curve in the near term.
3. Overall performance is a complex mix of the architectural design and the shading, lighting and HVAC systems that support them. The Times had the advantage of testing many aspects of this integration in the mockup prior to construction. Shading fabrics and control setpoints are always compromises between glare and view and daylight and interior design. Since not all occupants will respond the same way, the design provided for simple occupant overrides for the shades. The POE found that most shades ($\sim 80\%$) were rarely overridden but a smaller subset experienced many more overrides. Interestingly the overrides were often triggered by opposing responses—too much glare versus not enough daylight and view, reinforcing again the divergent nature of occupant preferences.
4. Active operator management is essential. Success requires a continuous commitment from the owner and AEC team from specification through all phases of design, to contracting and construction and finally with commissioning, occupancy, occupant training and maintenance. The Times prequalified the shading

system contractors with a training session before asking for bids and pre-assembled complex aspects of the lighting controls offsite to reduce errors and costs. Extensive commissioning activity was undertaken in every work area to ensure that systems worked as installed.

5. Engagement with occupants is crucial both for new technical systems and new work environments. Many occupants had worked in private offices and had never worked in open office areas. New occupants were informed of the expected operation of the shading and lighting systems and the interface between automation and their user control options. The owners tracked performance over time and contracted with the suppliers for support services over the early years to address any shortcomings that emerged.

5.7 Nordea Bank Headquarters

Anne Iversen, Micki Aaen Petersen and Jakob Strømann-Andersen
Henning Larsen Architects, Copenhagen, Denmark

Nordea Bank's new office building (Fig. 5.42) consists of two light sculptural volumes (seven floors) encompassing a total floor area of 40,000 m^2 and serving 1800–2200 employees. The aim of the building is to provide the best opportunities for all of Nordea's employees to work in an environment connected with daily and seasonal changes in daylight and views to the outdoors. The client additionally

Fig. 5.42 Rendered view of Nordea Bank Headquarters viewed from the south-west corner. *Image credit* Henning Larsen Architects

Table 5.6 Nordea Bank
Headquarters

Owner	Nordea Properties
Completion date	2016
Project type	Large commercial office building
Gross floor area	40,000 m^2
Architect	Henning Larsen Architects, Denmark
Engineer	COWI
Location	Ørestad, Copenhagen, Denmark

sought the project to serve as a model of sustainable building design. The Danish Building regulation requires building annual Energy Use Intensity (EUI) to be below 41 kWh/m^2-year. Relative to the national median EUI for existing large office buildings (71 kWh/m^2-year), meeting the EUI target amounts to a 58% energy reduction. In addition, Nordea now requires all new bank developments to achieve LEED Platinum, the highest score awarded by the international green building rating system (Table 5.6).

5.7.1 Integrated Daylighting Design

The project site is divided into four volumes, of which the two western volumes are part of Stage 1 of the Nordea Bank Headquarters building plan (Fig. 5.43). The volumes are placed on a base (level-01, 00), which creates a park landscape towards the south and appears more shielded by the two-story shale facade towards the north. On top of the base (level 01) the entire building footprint is used in order to create large, open spaces for the common facilities, the trading floor and the cafeteria. Atriums are placed in the center of the building mass and serve to spatially connect the first floor (level 01) to the upper floor (level 07) creating a feeling of unity between the various work zones within the large project. Daylight delivered through a three-dimensional grid of skylights above each atria serves to expose the core zones of the building to daily and seasonal changes in light as well as offer views of the sky from deep within the building from certain viewpoints while blocking direct view of the solar disc from work zones. Work zones are located around the atrium at the perimeter area of the building in order to use daylight as the primary means for ambient lighting. Secondary functions (such as meeting and break rooms) are located inward towards the two atria.

The greatest daylighting challenges were related to the 5500 m^2 trading floor (Fig. 5.44). The combination between Nordea's wishes, the project site and the technical criteria were quite contradictory. From a design perspective, it was Nordea's wish to have an open office with possibility of sky views. However, as the project site only allows a dense building geometry, there was no other choice than to have deep rooms within the building mass. This combination led to two

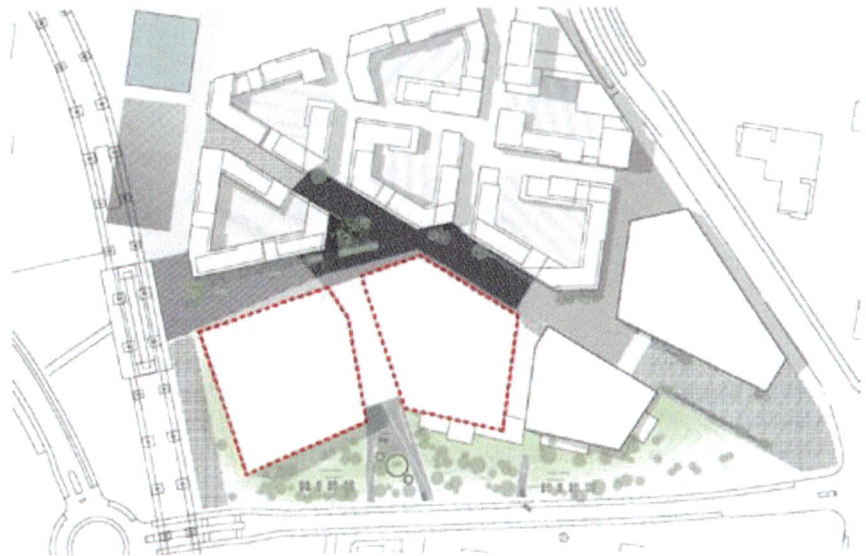

Fig. 5.43 Building plot for Nordea Bank Headquarters. *Image credit* Henning Larsen Architects

Fig. 5.44 Rendered view of trading floor atrium. *Image credit* Henning Larsen Architects

challenges. One, to provide the required daylight conditions and quality needed for work zones. Two, to avoid any direct daylight as all employees work with computer screens.

Daylight analysis, CIE overcast sky. Grid measured 850 mm above the floor. Distance between grid-points: 500 mm. Software: Ecotect Radiance. Reflectances: Floor 10%, ceiling 70%, interior walls 50%, window frames 50% adjacent buildings 20%. Image Credit: Henning Larsen Architects.

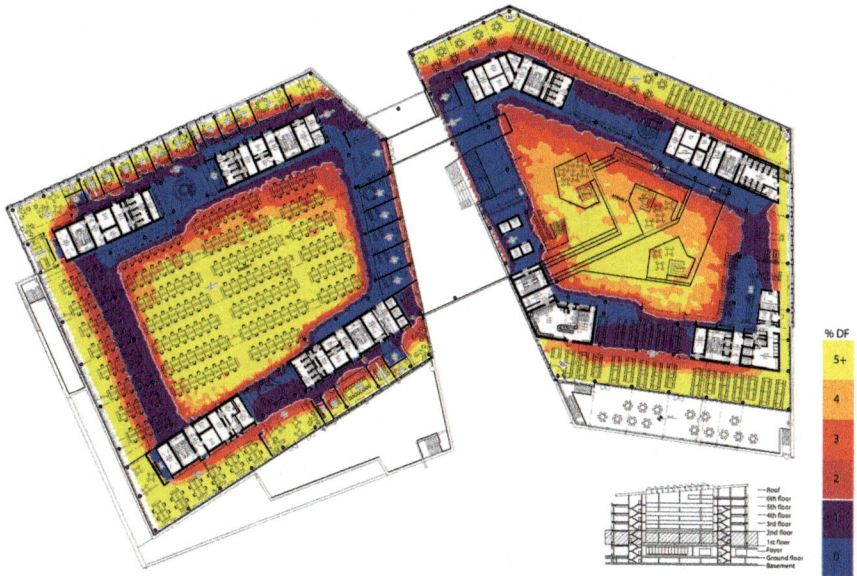

Fig. 5.45 Plan view showing daylight factor analysis for first floor office level of phase 1

The Danish building regulation requires that there should be either a minimum daylight factor (e.g. Fig. 5.45) in the permanent work spaces of 2% or a glass area in the facade corresponding to 10% of the floor area (or 7% if skylights are used). The project was additionally designed to comply with the LEED (v.3) Daylight and View Environmental Quality (EQ) credits. A final objective was to minimize the use of ambient electrical lighting during daylight hours by installing a daylight-dimming lighting control system.

5.7.2 Facade Systems

From an architectural perspective, the client desired the facade to appear visually transparent (Fig. 5.46). However from an energy perspective, it was a desire to minimize external solar gain as well as thermal heat losses. The design team addressed this challenge by developing a custom double-skin facade concept based on a Kastenfenster-facade (Fig. 5.47), a vertical box window concept that was originally developed in Germany. The facade concept is characterized by its modular construction, where a relatively large space (approximately 150 mm) between the inner and outer glass skins, creating a passively-ventilated air cavity in each element (Fig. 5.48). An automated exterior fabric roller shade is located within

Fig. 5.46 Rendering of the perimeter offices showing the level of openness and visual transparency of the facade. *Image credit* Henning Larsen Architects

this cavity that is deployed to control direct sun transmission and solar heat gains to the perimeter workspaces. Furthermore, the maintenance can be done easily from the inside of the building, where the modules can be opened manually. Interior automated fabric roller shades are located inboard of the facade to provide an additional layer of glare control for occupants.

The cavity plays a major role in energy and life-cycle performance of the facade by shielding the inner glass skin and exterior shading system from high wind forces (which would otherwise require the shading system to retract to avoid damage) as well as by reducing envelope convective heat losses. The cavity provides further benefits for indoor environmental quality by reducing transmission of external noise. Furthermore the modular system allows for customization within each element, enabling the flexibility in facade geometry and appearance required by the design team. Viewed from the interior (Fig. 5.46), the windows extend vertically from the floor to the ceiling to enable views to the outside both to the landscape, the city and the sky. A recess in the suspended ceiling allows transmitted daylight to penetrate deeper into the floor plan. Figures 5.49 and 5.50 show examples of the full scale mock-ups fabricated for evaluation prior to project construction.

The bank's safety policy requires use of laminated glass on both the inner and outer skins of the facade system. As a result, the system consists of a total of five (5) glass layers. The outer skin consists of a lamination of two layers of low-iron

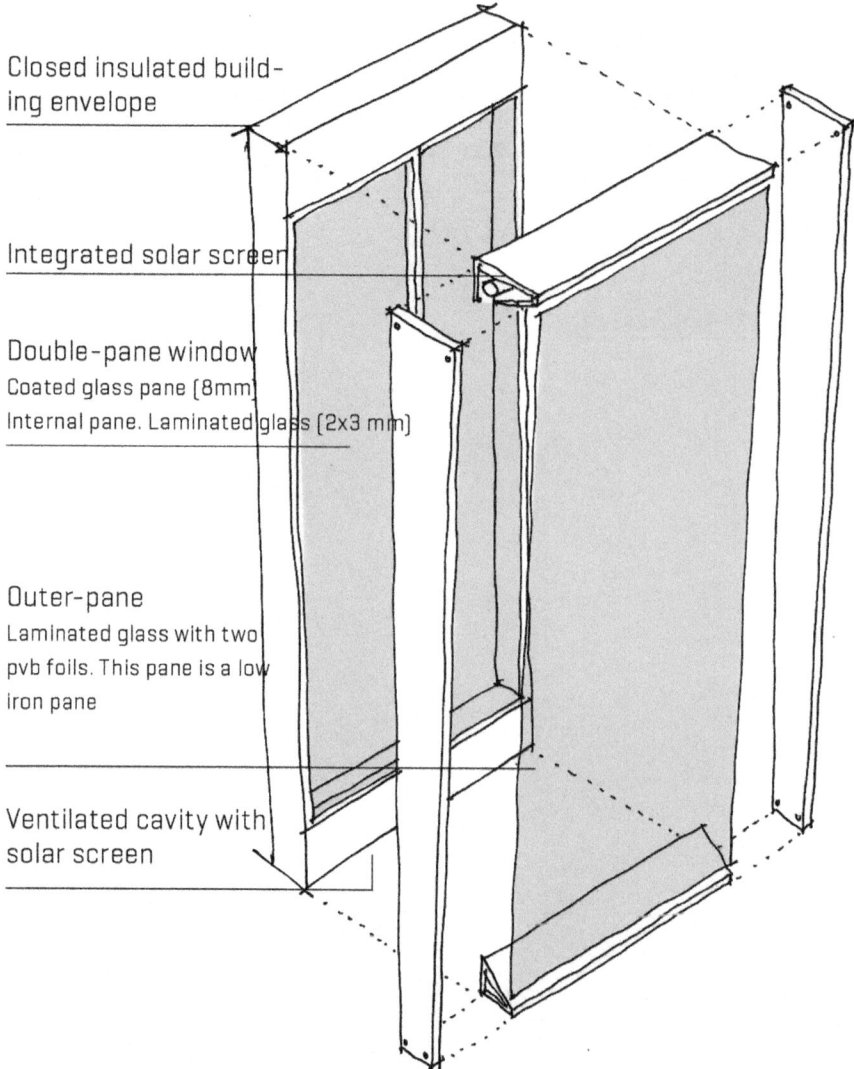

Closed insulated build-
ing envelope

Integrated solar screen

Double-pane window
Coated glass pane [8mm]
Internal pane. Laminated glass [2x3 mm]

Outer-pane
Laminated glass with two
pvb foils. This pane is a low
iron pane

Ventilated cavity with
solar screen

Fig. 5.47 Exploded view of kastenfenster-facade. *Image credit* Henning Larsen Architects

glass with a spectrally selective solar control coating on surface 2. The inner skin consists of an insulated glass assembly with an additional layer adjacent to the facade cavity. Due to the large number of total glass layers (5), care was taken by the design team to specify low-iron glass to maximize transmission of visible light through the glazing assembly, resulting in an overall VLT of 62%.

To extract the benefits of the achieved natural lighting level, the lighting control strategy uses closed-loop daylight-dimming, with the possibilities for the occupants

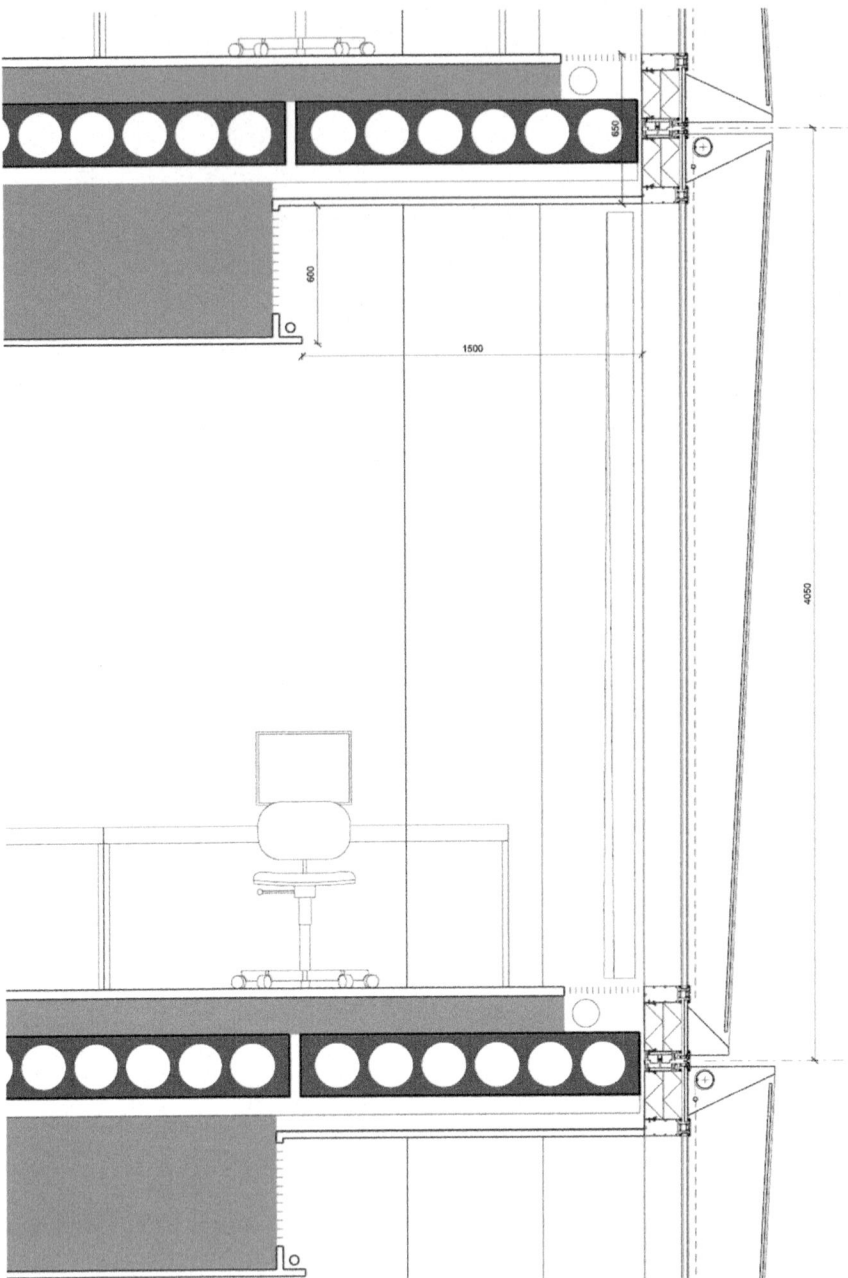

Fig. 5.48 Wall section showing integration of kastenfenster-facade with ceiling soffit to improve daylight penetration. *Image credit* Henning Larsen Architects

Fig. 5.49 Full scale mock-up. *Image credit* Henning Larsen Architects

to manually override the control settings. The photo-sensors are placed every 4 m^2 corresponding to the electrical lighting zones. The combination of an ambient lighting power density at 200 lx on 5.3 W/m^2 and the daylight daylight-dimming system enable electrical lighting energy consumption and corresponding internal loads to be reduced.

For the atrium design, the objective was to design a geometry that would allow a large amount of diffuse radiation. To minimize the amount of direct sunlight at trading floor, the skylight openings have been angled towards north (see Fig. 5.51). The intense use of computer screens at trading floor has led to far-reaching studies of the direct sun penetration through the skylights. The optimal design of the roof structure should screen for all direct sunlight. This would require a skylight geometry that is supplemented with relatively long fins with purpose of shielding the direct sun, even the low afternoon sun from west. However this design solution would unable the sky view criteria established by Nordea. Instead custom-made MicroShades^{TM} (see Chap. 3) are incorporated in the skylights, blocking the direct sun from the analyzed worse-case glare scenarios, while still letting the diffuse sky illuminance penetrate into the building and allowing the possibility for sky view.

Fig. 5.50 Mock up. *Image credit* Henning Larsen Architects

Fig. 5.51 Hourly rendered views used for study of direct sun penetration. *Image credit* Henning Larsen Architects

References

Anselmo F, Lauritano A (2003) Evaluation of the solar energy potential in urban settings by irradiation map production. Presentation at the radiance workshop in Berkeley California, 22–26 Sept 2003. https://www.radiance-online.org/community/workshops/2003-berkeley/presentations/Anselmo/radmap.pdf

City of Seattle Department of Construction & Inspections (2016) Living building pilot. http://www.seattle.gov/dpd/permits/greenbuildingincentives/livingbuildingpilot/. Accessed 2 Sept 2016

Government of Australia Commercial Building Disclosure Program. Building energy efficiency certificate. https://cbdportal.cbd.gov.au/Download/ShowPdf?id=B1800-2015-1. Accessed 2 Sept 2016

Guglielmetti R, Pless S, Torcellini P (2010) On the use of integrated daylighting and energy simulations to drive the design of a large net-zero energy office building. NREL/CP-550-47522. Presented at SimBuild 2010, New York, New York, 15–19 Aug 2010. http://www.nrel.gov/docs/fy10osti/47522.pdf

Guglielmetti R et al (2011) Energy use intensity and its influence on the integrated daylighting design of a large net zero energy office building. Presented at the ASHRAE winter conference, Las Vegas, Nevada, 29 Jan–2 Feb 2011

International Living Futures Institute (2012) Living building challenge (v2.1). http://living-future.org/sites/default/files/LBC/LBC_Documents/LBC%202_1%2012-0501.pdf. Accessed 2 Sept 2016

Lee ES, Fernandes LL, Coffey B, McNeil A, Clear R (2013) A post-occupancy monitored evaluation of the dimmable lighting, automated shading, and underfloor air distribution system in The New York Times Building. LBNL-6023E. https://buildings.lbl.gov/sites/all/files/lbnl-6023e.pdf

National Renewable Energy Laboratory (NREL) (2012) The design-build process for the research support facility. DOE/GO-102012-3293, June 2012. http://www.nrel.gov/docs/fy12osti/51387.pdf

Peña R (2014) Living proof, the Bullitt Center high performance case study. University of Washington Center for Integrated Design

Pless S, Torcellini P (2011) Using an energy performance based design-build process to procure a large scale low-energy building. NREL/CP-5500-51323. Presented at the ASHRAE winter conference, Las Vegas, Nevada, 29 Jan–2 Feb 2011. http://www.nrel.gov/docs/fy11osti/51323.pdf

Chapter 6
Validating Performance from the Perspective of End Users

Kyle Konis

6.1 Introduction

Effective daylighting requires rethinking the simplified approach to glazing and facade systems to acknowledge the needs and behaviors of building occupants as a critical determinant of long-term performance. Occupants represent a rich multi-sensory source of information on environmental performance. This chapter argues that a lack of human factors data from buildings in use leads to environmental design that is largely detached from the preferences or needs of building occupants, with cascading implications for occupant comfort and energy use. Emerging methods for collecting human factors data on Indoor Environmental Quality (IEQ) are presented and discussed for integrating detailed occupant-feedback into building evaluation, operation, and the design process. This chapter concludes by proposing an approach to environmental design informed by examination of occupant behavior, personal modifications, and subjective assessments of IEQ and speculates on how this approach may lead to better outcomes for building occupants.

6.2 Closing the Loop, Feedback and Learning

The ability to validate design assumptions requires effective methods and technologies to compare performance in use with design intent. Despite the development of increasingly sophisticated systems to measure building energy consumption, there remains a lack of effective tools for placing energy consumption in context with subjective assessments of end user comfort, or preferences for indoor environmental conditions. While real-time building energy monitoring and public disclosure of whole-building energy use are becoming increasingly common, buildings are rarely evaluated after occupancy, (referred to as Post Occupancy

© Springer International Publishing Switzerland 2017
K. Konis and S. Selkowitz, *Effective Daylighting with High-Performance Facades*, Green Energy and Technology, DOI 10.1007/978-3-319-39463-3_6

Evaluation (POE)), to understand building performance from the perspective of end users. In regard to daylighting, buildings are often promoted as successful examples of daylighting on the basis of having large areas of high Visible Light Transmittance (VLT) facade glazing, daylight-dimming lighting controls, or on the basis of compliance with the LEED Daylighting Environmental Quality (EQ) credit. Although buildings can be examined in use to compare measured physical environmental conditions against consensus-based assumptions for effective daylighting (e.g. IES LM-83) or LEED compliance requirements (e.g. LEED), there remains limited research from the field validating the applicability of these requirements for predicting the satisfaction of building occupants with the daylight sufficiency, view, visual and thermal comfort outcomes. And, as predictive methods become increasingly complex and dependent on hourly simulation, it is often unclear how to compare occupant subjective feedback from the field with annualized simulation results.

Further, to improve the design and performance of environmentally responsive technologies and architectural strategies, design teams need feedback that goes beyond a simple indicator of success or failure. Because environmentally responsive design strategies often result in indoor environmental conditions that are more dynamic than in conventional, mechanically controlled environments, design teams need feedback data with a sufficient level of spatial and temporal granularity to identify "when" and "where" assumptions for comfort and satisfaction are achieved and not achieved for a project in use. If acquired systematically, this data can be aggregated to begin to investigate and refine existing assumptions for comfort and satisfaction across a broad spectrum of climates, building types, site conditions, and uses.

To validate project performance and the design guidance on which projects are based, it is necessary to examine how the indoor environmental conditions enabled by a given design are assessed, modified (both formally and informally), and accommodated by building occupants over time and how these outcomes compare to design intent.

6.2.1   From Universal Design to Learned Guidance and Adaptive Systems

Occupant-centered data, when collected systematically across a range of environmental and contextual conditions, can be used in a number of ways to improve design outcomes. Data aggregated from multiple projects can be analyzed to establish unique data-driven performance goals for individual projects informed by factors such as climate, building type, program and user populations. These performance goals can serve as an evidence-based alternative to the application of universal standards or design "rules-of-thumb." Another use is to generate data-driven predictive models that can improve the fidelity of existing simulation-based design processes. Real-time and historic occupant data can also be used in the operation of dynamic facade and lighting systems to support algorithms which adjust ambient and local environmental conditions to meet the personal preferences of occupants rather than the requirements of a static, theoretical model. To achieve these possibilities, novel tools and techniques must first be developed to systematically assess indoor environments from the perspective of building occupants.

6.3 Adding Humans to the Loop—A User-Interface Design Problem

Understanding if buildings, in operation, are achieving performance expectations for IEQ is critical for wider adoption of green building practices. To achieve this goal requires effective tools for collecting feedback from occupants in buildings in use. Assessment of environmental conditions is particularly challenging in dynamic daylit environments. In spaces designed to achieve daylight autonomy, time of day, sun angle, exterior weather, and the position of movable shading devices are all dynamic factors that influence the intensity and distribution of available daylight. Consequently, acquiring physical or subjective measurements of daylight descriptive of the conditions experienced by occupants in buildings can be challenging. Addressing this challenge requires repeated engagement with occupants over daily and seasonal changes in solar and weather conditions. And, to validate indicators of daylight sufficiency or visual comfort with subjective data, subjective assessments must be collected simultaneously with physical measures. Finally, occupants may have different expectations for daylight availability based on their location in the building (e.g. depth-from-facade), therefore occupant spatial location must be considered.

6.3.1 From the Laboratory to the Field

Arguably, the challenges of isolating and controlling a single indoor environmental condition (i.e. "stimulus" variable) to investigate a human-factors outcome (i.e. "response" variable) in real work environments has served, historically, as the basis for reliance on controlled laboratory studies as the principal mode of investigation for human factors research on lighting (as well as other IEQ variables such as thermal comfort). While this approach may have been viewed as suitable for deriving IEQ criteria for sealed and mechanically conditioned buildings based on the assumption that conditions in these spaces were designed to be relatively steady-state and spatially-homogeneous, the approach is problematic when applied to assess the dynamic, often more extreme, and spatially-variable environmental conditions of passive and low-energy building designs. And, although laboratory studies enable a greater level of control, the applicability of results from laboratory studies are problematic when directly applied in the evaluation of real work environments. This is due to a number of discrepancies including differences in simulated vs. real work tasks, study duration, stress level, relative capacity of the study participant to modify and habituate to his/her environment, and the potential influence of context (e.g. quality of view) on subjective assessments. Therefore, due to the need to capture subjective assessments in real work environments, along with the challenges of replicating the dynamic indoor environmental conditions delivered by passive and low energy building designs, there is a strong incentive for human-factors research and technology development to move from the control of the laboratory setting to the more challenging context of real buildings in use.

6.3.2   Enabling Buildings as Living Laboratories

The objective of collecting occupant feedback from buildings in use is to allow all buildings to serve as living laboratories, enabling proof of performance for innovative designs and technologies, evaluated and improved with feedback from end-users. The desktop polling station (Fig. 6.1) is an interactive device developed to enable repeated measures of occupant subjective assessments paired with

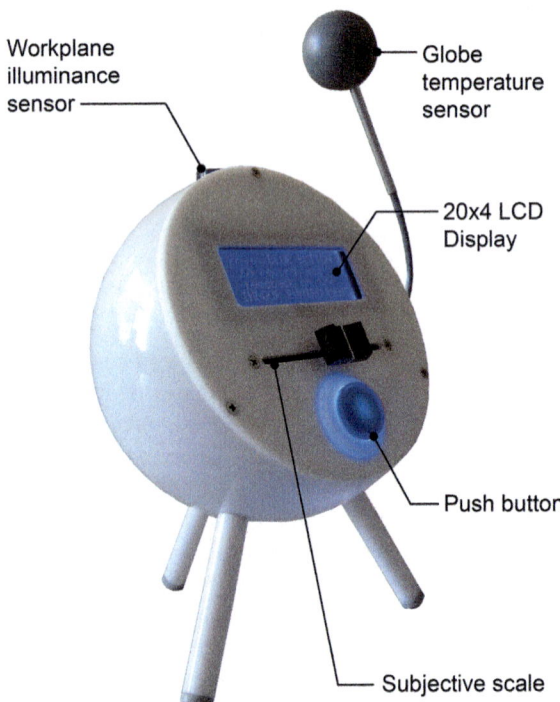

Fig. 6.1 Desktop polling station

Workplane illuminance sensor

Globe temperature sensor

20x4 LCD Display

Push button

Subjective scale

physical measurements of IEQ. The polling station was informed by similar devices developed to evaluate the potential effects of Demand Response (DR) precooling on occupant comfort (Xu and Zagreus 2006; Lee et al. 2007). The design of the device was also informed by field observations of work environments were many occupants were found to supplement their desktops with non-work-related objects which served to personalize the workspace. The design of the device as an addition to such a "desktop menagerie," rather than as an overt research instrument, follows the ubiquitous computing model defined as, "machines that fit the human environment instead of forcing humans to enter theirs" (York and Pendharkar 2004).

The desktop polling station is designed to serve as a non-disruptive interface to collect occupant feedback in the field (Fig. 6.2). Subjective data are input by responding to a short survey. The survey consists of multiple short questions that query occupant satisfaction level or preference at the point in time the survey is initiated (Fig. 6.3).

In addition to an interface for subjective measures, the polling station serves as a platform for physical sensors. The prototype (Fig. 6.1) includes a globe thermometer and global horizontal illuminance sensor. Globe temperature is measured using a globe thermometer attached to the side of the device. When in a state of equilibrium, a globe thermometer indicates the combined influence of radiative and convective heat exchange with a particular environment (e.g. a particular air temperature, air velocity, and temperatures of surrounding surfaces in an office)

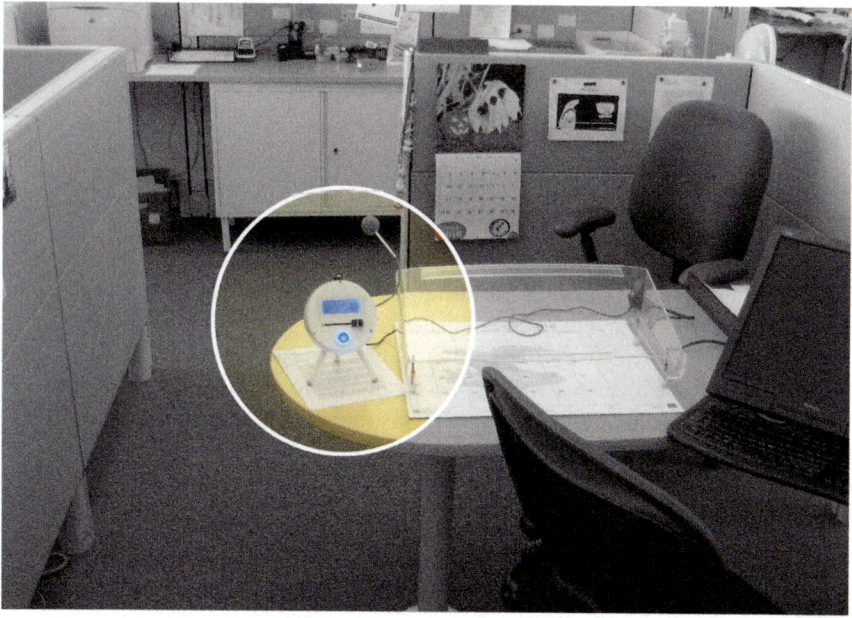

Fig. 6.2 Example deployment of a desktop polling station in the open-plan core zone of a daylit office building

Fig. 6.3 Example "point-in-time" occupant subjective assessment of daylight using the desktop polling station

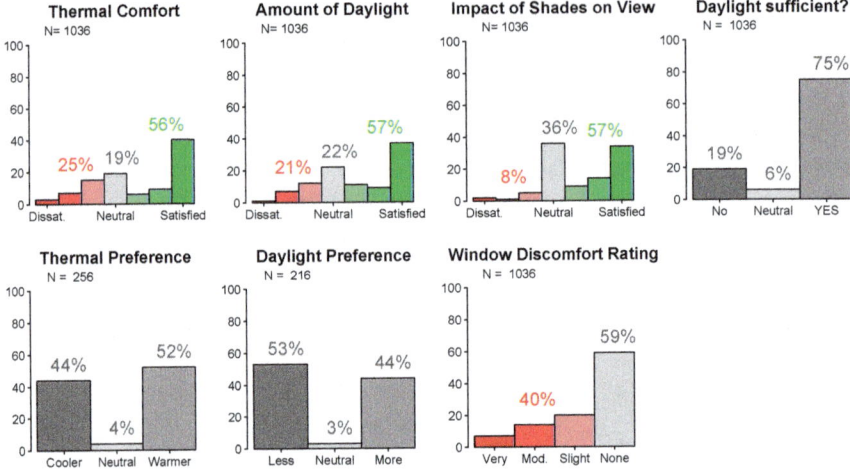

Fig. 6.4 Example "performance dashboard" view subjective responses to a seven-question repeated-measures survey instrument. Results are shown in aggregate for (N = 14) study participants

(Fountain 1987). The sensor used in the polling station is an epoxy encapsulated precision thermistor[1] suspended inside a spherical shell.[2] Global horizontal illuminance measures are made using a cosine-corrected photometric sensor.[3]

Detailed data provided by the polling stations can be analyzed for a range of different purposes. Figure 6.4 summarizes responses from (N = 14) participants to a seven-question repeated-measures survey instrument. Prior to examining data in context with physical lighting or temperature data, subjective data alone can serve as an indicator of overall performance. The "performance dashboard" visual format

---

[1]Brand = Measurement Specialties, type = 44016RC precision thermistor, resistance = 10,000 Ohms at 25 °C. Prior to assembly, thermistors were calibrated in a thermal bath to within +/− 0.1 °C.

[2]The shell is a ping pong ball spray painted 50% matt grey.

[3]Brand = Licor, type = LI-210, nominal accuracy = 3%.

shown in Fig. 6.4 can be used in early stages of analysis to examine how responses for each question were distributed across the subjective scale and to develop an overall understanding of how the building was performing from the perspective of occupants. Using this dashboard, problematic environmental factors can be identified and prioritized for further investigation, leveraging the spatial and temporal granularity of the data to determine "when" and "where" elements of the environmental design are performing successfully or unsuccessfully from the perspective of occupants.

6.3.3 Validating Daylighting Assumptions in Green Building Rating Systems

Occupant-centered data can also be analyzed to examine the applicability of existing assumptions for daylight sufficiency, such as those embedded in green building certification compliance criteria (e.g. LEED). The example below, shown in Fig. 6.5, presents data collected in the perimeter zone of a daylit office building (Konis 2011). Figure 6.5 compares subjective responses to the polling station question: "How satisfied are you with the amount of daylight in your workspace right now?" to the magnitude of daylight illuminance measured simultaneously at the polling station. The figure shows the distribution of all "satisfied" responses in green and all "dissatisfied" responses in red. N indicates the total number of participants for the monitoring phase followed by the total number of responses among all participants in parenthesis. Vertical lines are drawn to indicate threshold levels of 300 and 500 lx. The LEED Daylighting EQ credit uses a 300 lx horizontal illuminance threshold to determine if daylight levels are "sufficient" or not for

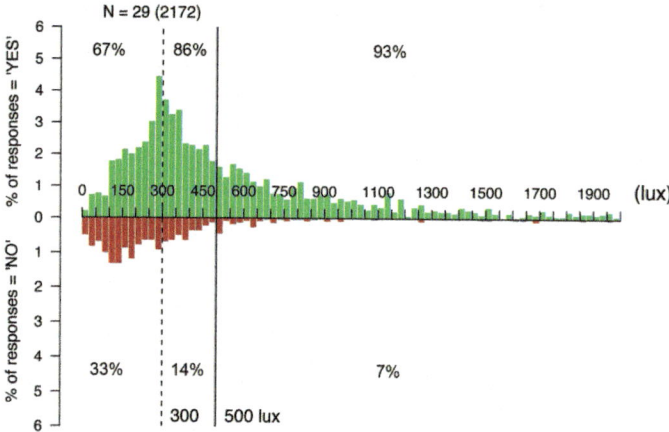

Fig. 6.5 Distribution of responses to satisfaction with amount of daylight

occupants. Thresholds of 300 and 500 lx are common recommendations for the minimum acceptable levels of workplane illumination for offices with computer-based and paper-based tasks respectively. The percentages of "YES" and "NO" responses for each subset of illuminance levels (0–300, 300–500, >500 lx) are shown on the figure. Figure 6.5 shows that the majority (67%) of responses recorded at workplane illuminance levels below 300 lx indicated the perception of sufficient daylight to work comfortably without supplemental electrical lighting and (86%) of responses at workplane illuminances of 300–500 lx.

Data from multiple polling stations, distributed to where occupants are located in the building, can be analyzed to generate evaluations of Spatial Daylight Autonomy (sDA) for comparison with simulation-based predictions. Successful performance based on sDA requires that a minimum of 75% of occupied spaces exceed the threshold illuminance of 300 lx for at least 55% of the analysis period. In Fig. 6.6, sDA was calculated by determining the percentage of the total polling station population where measurements exceeded the 300 lx threshold over the time interval of exterior daylight availability (nominally 6:00–18:00 PST). Vertical grey bars show the percentage of the polling station population (from 0 to 100%) that exceeded the DA illuminance threshold at 15-min intervals in aggregate over the 25-day analysis period. Dividing the number of intervals that exceeded the 75% spatial threshold by the interval total then leads to the sDA outcome. The zone achieves 75% Spatial Daylight Autonomy for only 12.5% of daylight hours.

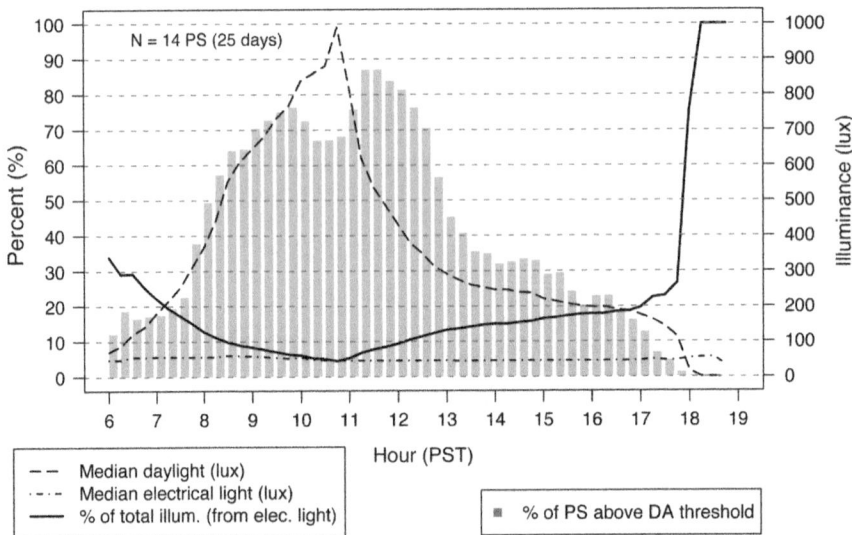

Fig. 6.6 Spatial daylight autonomy evaluation using data from 14 polling stations distributed to occupant workstations within the space being evaluated

6.3.4   Modeling Occupant Perception of Available Daylight

While many studies have found that occupants accept daylight levels outside of those recommended by existing recommendations, studies rarely model the data or demonstrate the factors that may affect spatial or seasonal changes in perception. In Fig. 6.7, logistic regression was used to examine responses to the survey question, "Could you work comfortably with the electrical lighting turned OFF right now? Responses were modeled in binary form (0 = "NO", 1 = "YES") in relation to concurrent measurements of horizontal workplane illuminance measured at the polling station. Figure 6.7 plots the probability of a "YES" response as a function of illuminance using data collected in two perimeter zones (NW facing, and SE facing) of a daylit office building over two multi-week monitoring periods for each zone. Vertical lines are drawn to indicate common threshold levels of 300, and 500 lx as well as an additional threshold of 100 lx. Because facade orientation and seasonal changes in solar position were considered confounding factors in the study, a separate model was applied to each phase of data. Data from (N = 29) unique participants were used, totaling 2422 unique responses. The models generated were found to correctly predict between 77 and 90% of observed responses.

Figure 6.7 shows a high probability that occupants perceive daylight to be sufficient at illuminance levels below the LEED Daylighting EQ 300 lx threshold criterion. Assumptions for the appropriate daylight sufficiency threshold are

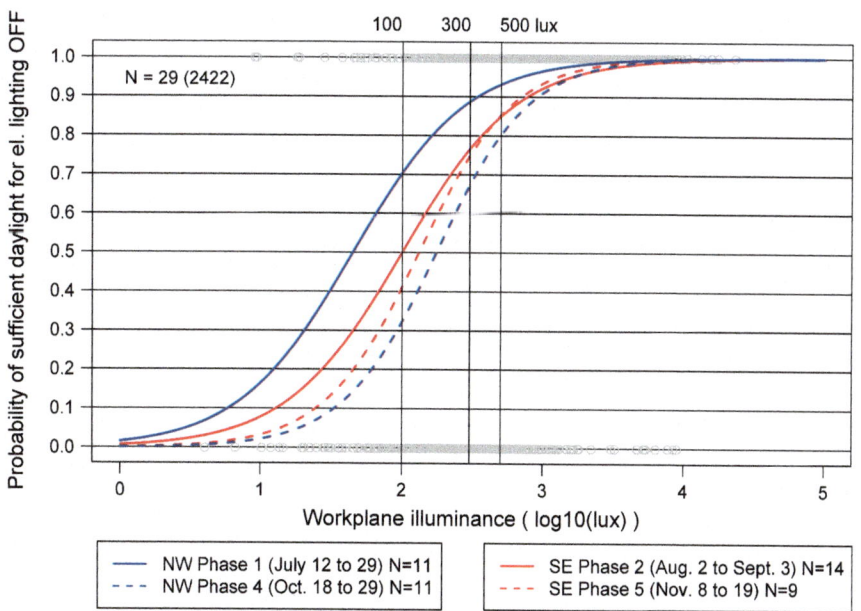

Fig. 6.7 Probabilistic model of occupant perception of daylight sufficiency

particularly important in design because they influence allowable floor plate depth, window glazing area, window glazing properties, and the configuration and operation of shading systems. In regard to energy, a daylight sufficiency threshold is used to determine when supplemental ambient electrical lighting should be turned on, or how much output should be dimmed by photocontrols. Figure 6.7 shows that in many instances, daylight was perceived to be sufficient to work comfortably without any electrical lighting at daylight levels below 100 lx. For example, the model based on data from the NW perimeter zone during Phase 1 (July 12–29) shows a 67% probability of sufficient daylight at a workplane illuminance of 100 lx, and an 89% probability at 300 lx. In addition, the variation between models suggests that facade orientation and seasonal changes in exterior solar conditions may influence perceptions of daylight sufficiency. For example, the probability of a "YES" response at 100 lx decreased for both NW and SE perimeter zone follow-up phases where seasonal changes resulted in fewer daylight hours and overall lower daylight levels due to more overcast sky conditions. A similar approach was used to compare subjective assessments of visual discomfort (e.g. glare) and thermal comfort to physical measures of luminance and globe temperature. This work is documented in Konis (2013, 2014).

6.3.5   Enabling Multi-sensory Investigation

Additionally, the technology enables multi-sensory investigation to be performed through cross-comparison between multiple subjective measures. As shown in Fig. 6.8, a statistically significant (p < 0.001) relationship (R^2 = 0.23) was found

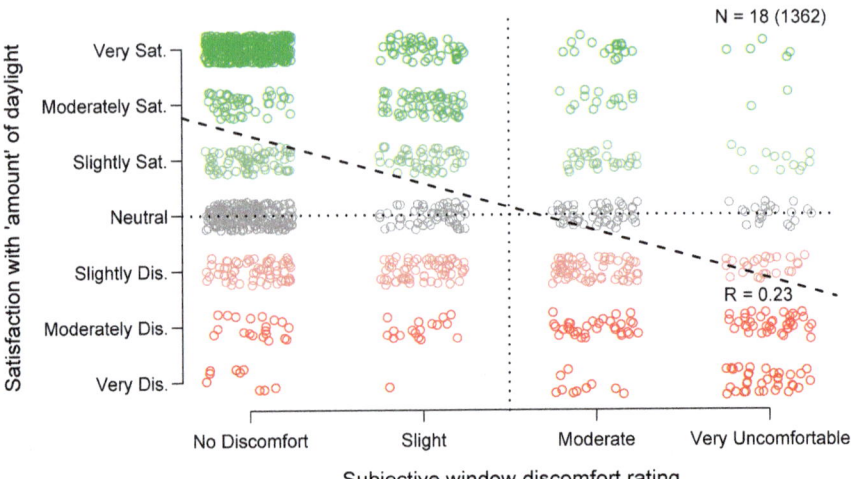

Fig. 6.8 Correlation of "satisfaction with amount" of daylight with subjective window visual discomfort rating

between satisfaction with the amount of daylight and window discomfort rating. The relationship shows that occupants are more likely to be satisfied with the amount of daylight in their workspace (regardless of the illuminance level) when the window view is visually comfortable. For example, nearly all responses of "very satisfied" with the amount of daylight are paired with responses of "no discomfort" from windows. This result suggests that building occupants consider visual discomfort when assessing their level of satisfaction with the magnitude of daylight illuminance in their workspace. The result also suggests that measures of horizontal workplane illuminance are insufficient as indicators of occupant subjective assessments of the "amount" of perceived daylight and that additional measures, such as vertical luminance maps, which can better-assess glare from windows are needed to effectively evaluate daylighting performance.

6.4 Scaling up Occupant-Centered Evaluation

The primary limitation of the desktop polling station approach is the limit of scale. Each device is a purpose-build sensor platform and must be physically distributed and collected, which limits its application to large numbers of buildings without substantial cost. To address this limitation, the approach was extended to leverage existing mobile devices and wifi infrastructure in commercial buildings as a platform for low-cost and scalable real-time data collection (Fig. 6.9).

The Occupant Mobile Gateway (OMG) is an iOS and Android client application (app) and server-side technology that transforms mobile devices into critical instruments for understanding and improving building performance from the perspective of end-users (Figs. 6.10 and 6.11). The OMG is the outcome of research examining the feasibility of leveraging the growing availability of sensors embedded in mobile devices paired with occupant subjective data and machine learning algorithms to automatically generate and communicate actionable information to designers, management staff, and other building stakeholders (Konis and Annavaram 2017). Client-side software interfaces with built-in (light), plug-in (temperature/RH), and networked (e.g. CO_2) sensors to put occupant subjective assessments in context with detailed measurements of IEQ.

The prototype technology significantly advances the ability of project teams to assess, learn and continually improve operations and design practices by placing detailed, real-time feedback in context with dynamic occupant spatial location. Knowing a user's location in near real time presents significant benefits for evaluating IEQ because it reveals the spatial location of subjective assessments and physical measurements, enabling these factors to be mapped over architectural plan drawings or input into digital Building Information Models (BIM). In addition, it enables variations in zone occupancy to be modeled and compared with static assumptions that are often used for building lighting and mechanical equipment scheduling.

Subjective data will increase in value towards informing building IEQ performance to the extent that they can be contextualized with supporting spatial,

Cloud
Occupant subjective feedback on Indoor Environmental Quality is collected using standardized and customizable survey instruments managed from a remote server.

Temperature
The software is designed to interface with the THERMO-DO temperature / relative humidity sensor, enabling personalized thermal conditions to be collected in real time.

Light
The prototype explores the re-purposing of the built-in light sensors to monitor transient changes in interior lighting conditions.

Air Quality
Bluetooth Low Energy (BLE) technology is utilized to network with remote physical sensors such as CO_2 to expand the sensing capabilities of the diagnostic tool.

Fig. 6.9 The occupant mobile gateway (OMG) App

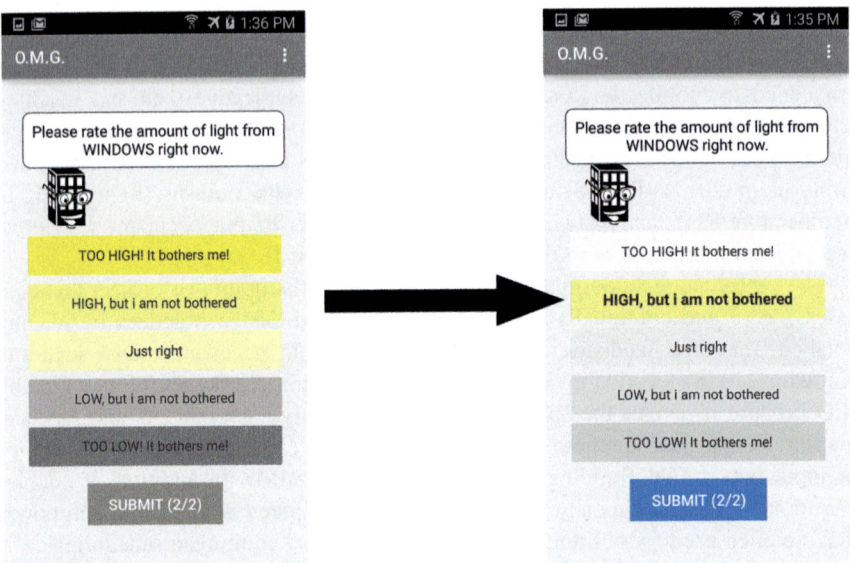

Fig. 6.10 Example "button style" interface showing initial state (*left*) and state after button press (*right*)

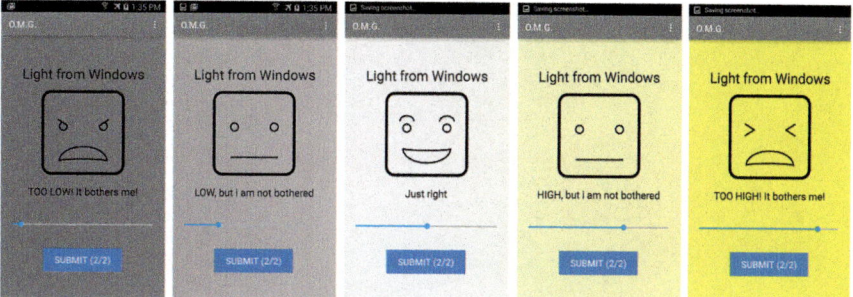

Fig. 6.11 Example "likert scale style" interface

temporal, and environmental data. While the location of mobile devices can be accurately determined in outdoor applications using Global Positioning System (GPS) data, GPS is not useful for indoor applications due to the interference of the building enclosure. The wifi mapping approach used by the OMG technology consists of two phases, an offline configuration phase, and an online operational phase. During the configuration phase, a map of the environment is created by collecting a set of all Access Point (AP) names and corresponding Received Signal Strength (RSS) measurements at coordinate-mapped locations across the floor plan. Figure 6.12 shows the location and signal strength of each AP for the environment used to validate the approach. The result is a database containing a unique "fingerprint" for each location. During the operational phase, the RSS measurements recorded from a given device are matched against the fingerprint dataset to find the closest match. The coordinates of the closest match from the map are then used to locate the device. This process occurs at regular (e.g. 1-min) intervals during the operational phase for all devices running the OMG application. Results from field-validation in a large, open plan workspace demonstrate that indoor location can be resolved at "desk-level" spatial resolution (Fig. 6.13).

The OMG technology provides occupants with a greater level of input on the management of their personal environment and establishes a systematic channel for providing feedback on IEQ. By collecting detailed physical/subjective IEQ data across multiple projects, the OMG technology has the potential to help enable evidence-based guidance for the architecture, engineering and construction community. This guidance can be applied in the development of more energy efficient, granular and responsive environmental control strategies in line with achieving Zero-Net-Energy (ZNE) building performance goals. With sufficient scaling, guidance can also be applied to validate and refine the static models of occupant comfort and lighting needs in current energy standards and green building compliance criteria.

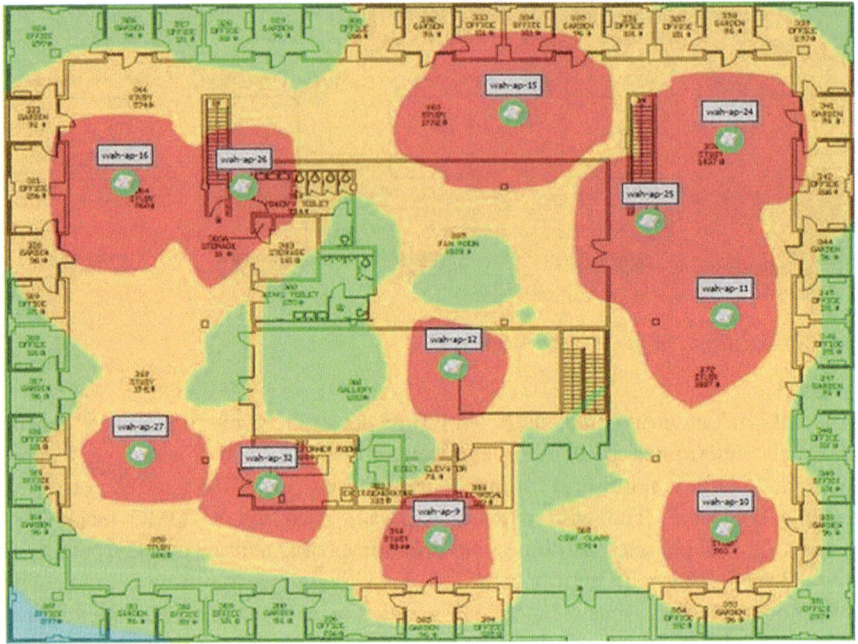

Fig. 6.12 Wifi signal strength map of one floor of a large academic building with perimeter cellular offices and internal open-plan workspace

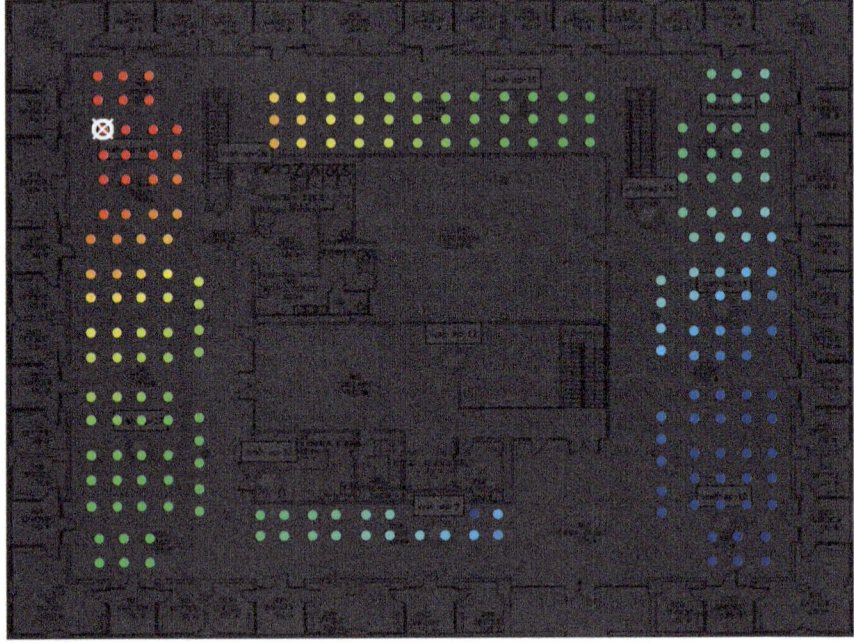

Fig. 6.13 Predicted location of sensed data. *Dots* show desk location. *Falsecolor mapping* shows likelihood of match

6.5 Learning from Occupant Shade Use and Personal Modifications

Because occupant use of interior shading devices can significantly reduce the level of daylight transmission as well as visual connection to the outdoors, it is important to understand how shading devices are controlled by occupants in daylit buildings and how shade use impacts daylighting, energy and IEQ objectives. And, to effectively predict daylight availability during design, it is important to translate knowledge from observations of buildings in use to empirical models that can be applied to predict occupant control of shading devices. Field data describing occupant control of shading devices is extremely limited, and existing assumptions used to predict the deployment of shading devices and frequency of operation vary widely. In addition, existing approaches to modeling shading devices assume simplistic facade configurations typical of conventional office buildings, which are ambiguous when applied to many sidelighting strategies, such as facades subdivided into a lower (view) zone and upper (daylight) zone due to the more complex shading configurations available to occupants.

Beyond formal controls available to occupants to address unwanted glare and solar heat, it is not surprising that occupants often make informal modifications to their work environments if the architectural design is not capable of delivering acceptable (or preferred) environmental conditions. For example, from observations collected in a POE of the San Francisco Federal Building (SFFB) by one of the authors (Konis 2011), a range of informal modifications to personal workspaces were observed that were initiated by building occupants as a means of supplementing available environmental controls provided by the design and subsequent facade shading retrofits. A collection of photographed examples is presented in Fig. 6.14. Each row shows three examples of a distinct approach to achieve a greater level of environmental control.

Several lessons can be learned regarding project performance from observation of these examples. The upper row presents informal modifications to add an additional fixed layer of solar and glare control at the facade. The second row shows supplemental task lighting and examples of modified built-in task lighting switched on during daylight hours to increase lighting levels adjacent to facade glazing that had been shaded by roller shades. Task lights built into workstations have been covered with translucent office paper to reduce glare from direct view of fluorescent bulbs. The second row illustrates the challenge of balancing energy objectives (for reduced electrical lighting) with occupant comfort and acceptable visual conditions for work tasks. The third row presents three examples of modifications observed from workspaces were occupants preferred working without roller shades. In each example, occupants chose to address solar and glare conditions through local adjustments at the body (e.g. sunglasses, baseball hat) and at the visual task (computer screen) rather than at the scale of the entire workspace. Although these modifications represent examples were the original design intent for an unshaded, daylit perimeter zone were achieved, this outcome was observed for only a small

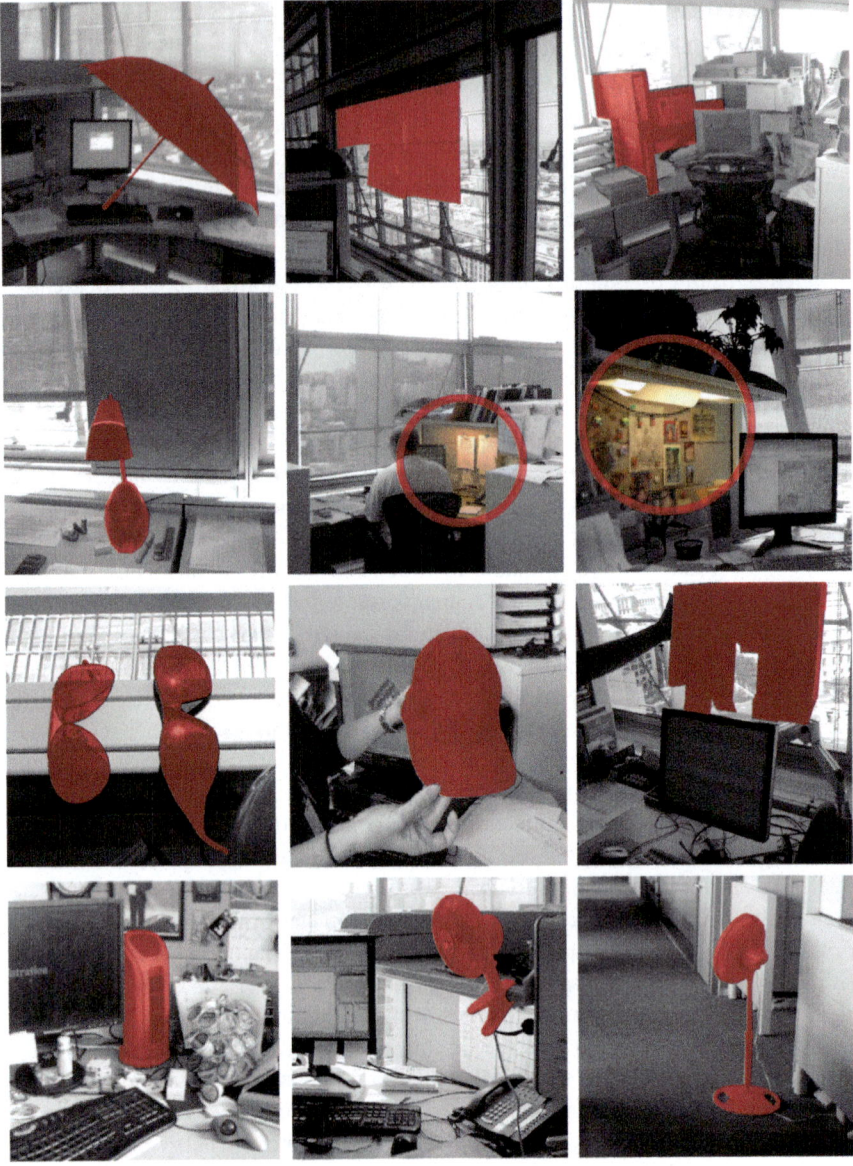

Fig. 6.14 Examples of occupant modifications to personal workspaces in response to indoor environmental conditions

fraction of workspaces in the study. The fourth row, similar to the second, shows examples of occupant efforts to supplement the heating and cooling provided by the building with portable heaters and fans and represents an additional energy demand not considered in the original design energy concept. Beyond simply indicators of project performance, thoughtful designers can interpret these examples as a basis for new design approaches that seek a more holistic integration of facade systems, interiors, ambient and task lighting and space conditioning systems, and occupant controls.

6.6   The Value of User Interfaces in Environmentally Responsive Architecture

As designers respond to the imperative to shift from resource-intensive modes of environmental control to modes that integrate environmental services provided by natural systems, points of interface for local environmental control become a central concern. Points of interface, in this context, refer to personal devices (e.g. operable windows/vents, interior/exterior shading and glare-control devices, task/ambient lighting, fans, etc.) that mediate between the occupant and the environmental conditions enabled by the building. In addition to simply providing these interfaces to building occupants, effective environmental design requires designers to investigate the performative aspects of these interfaces as well as the baseline conditions established by the building design. The retrofitting of promising daylit and environmentally responsive buildings such as the SFFB with solar control film and roller shades (Konis 2011) illustrates the challenge facing designers for balancing environmental design objectives with occupant comfort. Even after operable shades were added, the predominantly shaded configurations and low frequency of use observed demonstrate that simply providing controls to enable occupants to modulate environmental conditions does not necessarily result in their dynamic use, or in satisfied occupants. Moreover, designers must also consider the interaction between multiple controls. For example, the predominantly lowered shading devices and ad-hoc modifications to the facade (e.g. cardboard, umbrellas and other shading devices) served to restrict access to the manual controls for operable windows, reducing the effectiveness of natural ventilation and leading to increased frequency of electrical task lighting and fan use during the day. In shared environments, conditions must also be adjustable with sufficient granularity to enable occupants in shared spaces to adjust conditions locally without negatively influencing their neighbors. In addition to individual customization for comfort, granular control also has the potential to yield significant energy benefits. For example although the ambient daylight levels were considered sufficient by occupants to work comfortably in the perimeter zones without electrical lighting, the ambient overhead electrical lighting was never switched off, despite available wall controls. This outcome was due, largely, to the scale and autonomous control of the lighting system.

6.7   Conclusions

Closer consideration of occupant experience in buildings is integral to meeting the need for resource-efficient and climate-resilient buildings. In the absence of reliable physical performance measures for indoor environmental quality, scalable field-based methods are needed to acquire feedback directly from end-users to effectively evaluate performance and differentiate effective environmental design practices and technologies. Detailed occupant feedback data will increase in value towards improving the comfort and energy performance of buildings to the extent that data can be utilized to inform the design of environments that enable local user-control and support user modifications and adjustments.

To achieve better outcomes for occupant comfort and energy requires a flexible approach to design informed by lessons from built projects in use as well as more flexibility in the built projects themselves. By providing effective interfaces for occupants to make local environmental adjustments, designers can enable occupants as active participants in determining the conditions of their environment. Rather than passive recipients of indoor environmental conditions, occupants represent a rich multi-sensory source of information on environmental performance with the potential to serve as vital resource to better understand and respond to the complex relationship between the built environment and its inhabitants. Subjective data will increase in value towards informing building IEQ performance to the extent that they can be contextualized with supporting spatial, temporal, and environmental data.

Utilizing occupant feedback is not just about reducing energy, it is also about creating comfortable and desirable environmental conditions that support healthy and productive modes of use. Merging social science research methods with diagnostic technologies aimed and understanding user experience, particularly comfort and preferences related to indoor environmental conditions, can provide the basis for testing and refining design assumptions and serve as an empirical basis for modifying existing design standards (e.g. ASHRAE, IESNA) that often mandate extensive mechanical services. Rethinking engineering assumptions for human comfort and exploring alternate, less energy-intensive, approaches will become an increasingly relevant task as large developing economies mimic conventional U.S. building designs and environmental control approaches. Moreover, detailed occupant feedback, collected from buildings in use, can be used to validate and refine environmentally responsive design strategies, rewarding thoughtful design over application of standard practice or engineering "rules of thumb."

References

Fountain M (1987). Accuracy versus response time. Choosing a globe thermometer for thermal comfort measurements. Technical report to the U.C. Berkeley Center for Environmental Design Research (CEDR)

Konis K (2011) Effective daylighting: evaluating daylighting performance in the San Francisco Federal Building from the perspective of building occupants. Doctoral Dissertation, University of California, Berkeley, 420 pp

Konis K (2013) The influence of occupant behavior on facade solar transmission. Published in the proceedings of the ASHRAE summer conference in Denver, CO, 22–26 June 2013

Konis K (2014) Predicting visual comfort in side-lit open-plan core zones: results of a field study pairing high dynamic range images with subjective responses. Energy Build 77:67–79

Konis K, Annavaram M (2017) The occupant mobile gateway: a participatory sensing and machine-learning approach for occupant-aware energy management. Build Environ 118:1–13. doi:http://dx.doi.org/10.1016/j.buildenv.2017.03.025

Lee K, Braun J, Fredrickson S, Konis K, Arens E (2007) Testing of peak demand-limiting using thermal mass at a small commercial building. Demand Response Research Center Report, LBNL, July, 38 pp

Xu P, Zagreus L (2006) Demand shifting with thermal mass in large commercial buildings. LBNL-61172

York J, Pendharkar PC (2004) Human–computer interaction issues for mobile computing in a variable work context. Int J Hum Comput Stud 60:771–797

Printed by Printforce, the Netherlands